FOR 8 GRADE
TEACHER RECOMMENDED

Kids SUMMER ACADEMY

ARGOPREP

7 DAYS A WEEK
12 WEEKS

- Mathematics
- English
- Science
- Reading
- Writing
- Experiments
- Mazes
- Puzzles
- Fitness

GRADE 7-8

ArgoPrep is one of the leading providers of supplemental educational products and services. We offer affordable and effective test prep solutions to educators, parents and students. Learning should be fun and easy! To access more resources visit us at www.argoprep.com.

Our goal is to make your life easier, so let us know how we can help you by e-mailing us at: info@argoprep.com.

ISBN: 9781946755933
Published by Argo Brothers.

- ArgoPrep is a recipient of the prestigious **Mom's Choice Award**.
- ArgoPrep also received the 2019 **Seal of Approval** from Homeschool.com for our award-winning workbooks.
- ArgoPrep was awarded the 2019 **National Parenting Products Award, Gold Medal Parent's Choice Award** and **the Tillywig Brain Child Award.**

TABLE OF CONTENTS

TABLE OF CONTENTS

TABLE OF CONTENTS

TABLE OF CONTENTS

HOW TO USE THE BOOK

Welcome to **Kids Summer Academy** by ArgoPrep.

This workbook is designed to prepare students over the summer to get ready for **Grade 8.**
The curriculum has been divided into **twelve weeks** so students can complete this entire workbook over the summer.

Our workbook has been carefully designed and **crafted by licensed teachers** to give students an incredible learning experience.
Students start off the week with English activities followed by Math practice. Throughout the week, students have several fitness activities to complete. Making sure students stay active is just as important as practicing mathematics.
We introduce yoga and other basic fitness activities that any student can complete. Each week includes a science experiment which sparks creativity and allows students to visually understand the concepts. On the last day of each week, students will work on a fun puzzle.

HOW TO WATCH VIDEO EXPLANATIONS

IT IS ABSOLUTELY FREE

Go to **argoprep.com/summer8**
OR scan the QR Code:

WHAT TO READ OVER THE SUMMER

One of the best ways to increase your reading comprehension level is to read a book for at least 20 minutes a day. We strongly encourage students to read several books throughout the summer. Below you will find a recommended summer reading list that we have compiled for students entering into Grade 8 or simply visit us at: argoprep.com/**summerlist**

Author: J. D. Salinger
Title: Catcher in the Rye

Author: Michelle Obama
Title: Becoming

Author: Douglas Adams
Title: The Hitchhiker's Guide to the Galaxy

Author: Isaac Asimov
Title: Foundation

Author: Joan Bauer
Title: Hope Was Here

Author: Robert Cormier
Title: The Chocolate War

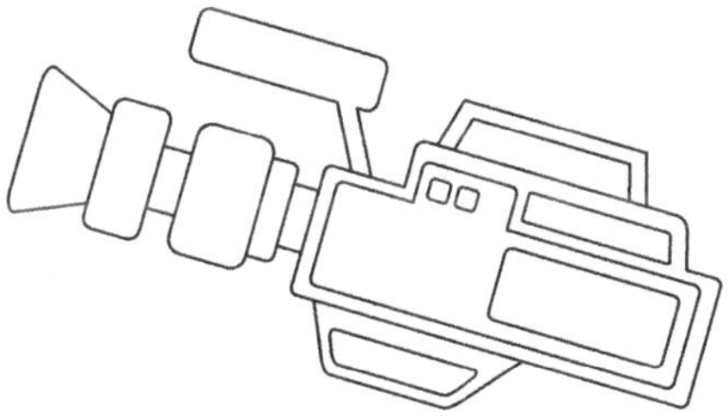

Author: Charles Dickens
Title: A Christmas Carol

Author: Alexandre Dumas
Title: The Three Musketeers

Author: Russell Freeman
Title: We Will Not Be Silent

Author: Anita Lobel
Title: No Pretty Pictures

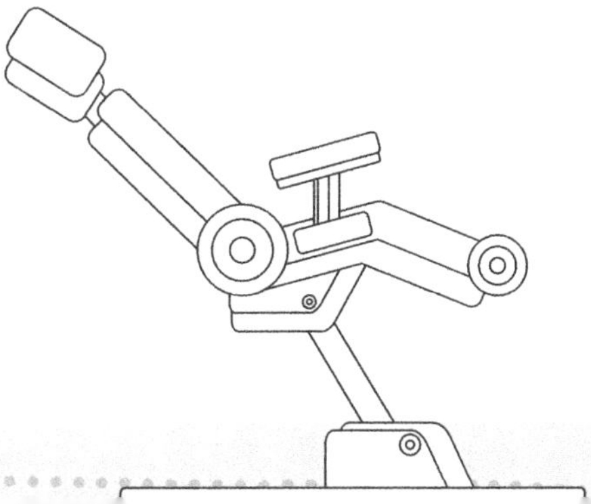

OTHER BOOKS BY ARGOPREP

Here are some other test prep workbooks by ArgoPrep you may be interested in. All of our workbooks come equipped with detailed video explanations to make your learning experience a breeze! Visit us at www.argoprep.com

COMMON CORE MATH SERIES

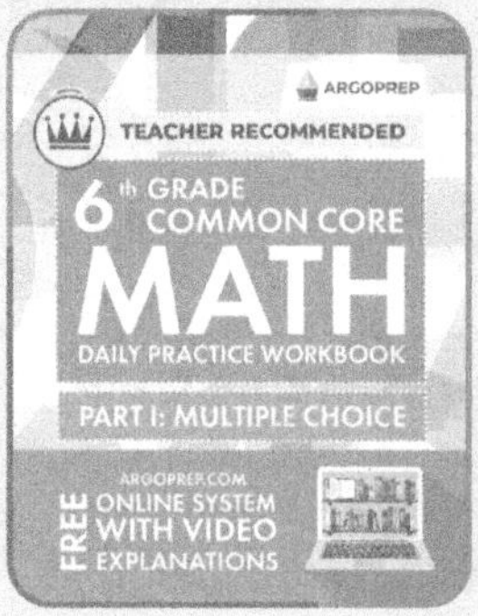

COMMON CORE ELA SERIES

KIDS SUMMER ACADEMY SERIES

ArgoPrep's **Kids Summer Academy** series helps prevent summer learning loss and gets students ready for their new school year by reinforcing core foundations in math, english and science. Our workbooks also introduce new concepts so students can get a head start and be on top of their game for the new school year!

Give your character a name

____

Write down the special ability or powers your character has and how you will help your community with the powers.

____

____

Great! You are all set. To become an incredible hero, we need to strengthen our skills in **English, math,** and **science**. Let's get started.

WEEK 1

ARGOPREP

WEEK 1 DAY 1

OVERVIEW OF ENGLISH CONCEPTS

Reading with Genre in Mind (Fiction)

You're probably already familiar with the concept of literary **genres**, but this week we'll dig a little deeper and think about how you can use your understanding of genre to make you a more effective reader <u>and</u> writer! When you know the **genre** of a literary work you're going to <u>read</u>, you have a sense of what to expect from it in terms of structure and theme. When you set out to <u>write</u> a story, a knowledge of the genre you're working in **simplifies planning** the story and gives you reliable **tropes** that your readers will recognize (more on that in a minute!).

Key Terms:

Literary Genre: A category or grouping of texts that share common themes, settings, structures, or techniques
Trope: A commonly used <u>plot point, literary device, symbol</u> or <u>character type</u> within a certain genre

Some Examples of Literary Genres & Tropes:

- **Realistic Fiction:** Fiction that takes place in "the real world," where the setting of the story is largely recognizable or relatable to the reader. (**TROPES:** The struggles of everyday life [getting gas, getting lost while driving, dealing with difficult people, etc.])
- **Fantasy:** Fiction that is not bound by the limits of what is possible in "the real world." (**TROPES:** Magic, fantastic or imaginary creatures, quests, magical swords and armor, etc.)
- **Science Fiction:** Stories based around the social, cultural, and psychological implications of technology, innovation, and progress. (**TROPES:** Robots, powerful computers, aliens, spaceships, other planets, mad scientists, etc.)
- **Mystery:** Stories that focus on investigations into criminal, strange, or mysterious behavior with an emphasis on figuring out what happened. (**TROPES:** police, lying witnesses, private detectives, thefts, murder, spies, etc.)
- **Historical Fiction:** Fictional stories set in specific times and places from true history, sometimes involving real historical figures. (**TROPES:** war, travel, ancient culture, important historical figures, key moments in history, etc.)
- **Horror:** Fictional stories designed to simulate scary or intense situations for the reader. (**TROPES:** monsters, violence, paranoia, fear, people being trapped, etc.)
- **Drama:** Texts designed to be performed on a stage rather than simply read. (**TROPES:** The tropes of a drama will vary based on its **subgenre**: Is it a <u>mystery</u> play? Is it a <u>comedy</u> play? Is it a <u>fantasy</u> play? Etc.)
- Of course, there are **many other literary genres out there**, but these are some of the most common. Try to think of what other genres you know!

WEEK 1 DAY 1 OVERVIEW OF ENGLISH CONCEPTS

Making Genre Study Part of Your Reading Habits:

- **Before you start a new book**, think about its genre. Read the cover or jacket, and write down a few predictions in your notebook or journal about what you think might happen in the story based on what you know about the **genre** and the most basic possible summary of what the book is about.
    - People say "**you can't judge a book by its cover**," but with an understanding of genre, you can at least make an educated guess!
    - Referring back to these predictions at various points as you read and especially after you read can provide opportunities for reflection about how the story either meets or defies expectations.
- Keep notes on the **use and development of tropes** as you read.
    - One of the things that can help you see which authors are the best (or at least your favorite) at writing within your favorite genre is to notice how each text **uses common tropes**.
        - How do they balance being **original and new** with using reliable aspects of the genre that everybody knows and understands?
- Evaluate the text **not just on its own, but as a representation of its genre**.
        - Does the author just rely on **tropes** in conventional ways that readers expect, or does the author try to provide **new spins** or defy expectations?
            - Does the story fit neatly within its genre, or is it **pushing the limits** in interesting ways?

From "Myths and Legends of Ancient Greece and Rome"

By E.M. Berens

The ancient Greeks had several different theories with regard to the origin of the world, but the generally accepted notion was that before this world came into existence, there was in its place a confused mass of shapeless elements called Chaos. These elements becoming at length consolidated (by what means does not appear), resolved themselves into two widely different substances, the lighter portion of which, soaring on high, formed the sky or firmament, and constituted itself into a vast, overarching vault, which protected the firm and solid mass beneath.

Thus came into being the two first great primeval deities of the Greeks, Uranus and Ge or Gaea.

Uranus, the more refined deity, represented the light and air of heaven, possessing the distinguishing qualities of light, heat, purity, and omnipresence, whilst Gaea, the firm, flat, life-sustaining earth, was worshipped as the great all-nourishing mother. Her many titles refer to her more or less in this character, and she appears to have been universally revered among the Greeks, there being scarcely a city in Greece which did not contain a temple erected in her honour; indeed Gaea was held in such veneration that her name was always invoked whenever the gods took a solemn oath, made an emphatic declaration, or implored assistance.

Uranus, the heaven, was believed to have united himself in marriage with Gaea, the earth; and a moment's reflection will show what a truly poetical, and also what a logical idea this was; for, taken in a figurative sense, this union actually does exist. The smiles of heaven produce the flowers of earth, whereas his long-continued frowns exercise so depressing an influence upon his loving partner, that she no longer decks herself in bright and festive robes, but responds with ready sympathy to his melancholy mood.

The first-born child of Uranus and Gaea was Oceanus, the ocean stream, that vast expanse of ever-flowing water which encircled the earth. Here we meet with another logical though fanciful conclusion, which a very slight knowledge of the workings of nature proves to have been just and true. The ocean is formed from the rains which descend from heaven and the streams which flow from earth. By making Oceanus therefore the offspring of Uranus and Gaea, the ancients, if we take this notion in its literal sense, merely assert that the ocean is produced by the combined influence of heaven and earth, whilst at the same time their fervid and poetical imagination led them to see in this, as in all manifestations of the powers of nature, an actual, tangible divinity.

READING PASSAGE

1. How does this text use **personification**?

2. In Paragraph 4, what does the author mean when he writes, "taken in a figurative sense, this union actually does exist" between the earth and the heavens?

3. According to the text, which person or thing existed **first**?

   A. Uranus
   B. Gaea
   C. Chaos
   D. Oceanus

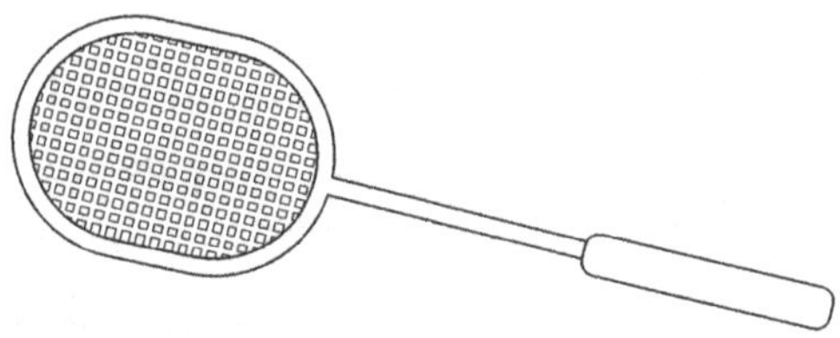

4. Which of these is the nearest synonym for "**primeval**," as it is used in Paragraph 2?

   A. Wicked
   B. Creative
   C. Most Important
   D. Ancient

5. **Based on the passage**, what did the ancient Greeks who created this myth believe the Earth looked or was shaped like?

WEEK 1 DAY 1

OVERVIEW OF ENGLISH CONCEPTS
ACTIVITIES

Recognizing Genres & Making Predictions Accordingly

Directions: Each sentence below provides a summary of a **literary text**. Read the summary and identify which of the **seven different genres mentioned in the concept overview** (realistic fiction, fantasy, science fiction, mystery, historical fiction, horror, or drama) you believe that story would probably belong to. Once you've identified the genre, **make a prediction** about another plot detail, character, or **trope** that you expect will be in the book based on its genre and what you've read in the summary.

1. **SUMMARY:** A brother and sister raise money to help pay for their grandmother's cancer treatment.

    a. **GENRE:** ____________________

    b. **MAKE A PREDICTION** (name at least one thing you would expect from this story - a character, moment, or trope) based on the summary you read and your understanding of **genre:** ____________________

    ____

    ____

2. **SUMMARY:** A group of brave knights must come together to fight a witch who lives in a far-away castle.

    a. **GENRE:** ____________________

    b. **MAKE A PREDICTION** (name at least one thing you would expect from this story - a character, moment, or trope) based on the summary you read and your understanding of **genre:** ____________________

    ____

    ____

WEEK 1 DAY 1

OVERVIEW OF ENGLISH CONCEPTS
ACTIVITIES

3. **SUMMARY:** Nobody can figure out who stole the most valuable jewel in Europe.

   a. **GENRE:** ______________________________

   b. **MAKE A PREDICTION** (name at least one thing you would expect from this story - a character, moment, or trope) based on the summary you read and your understanding of **genre:** ______________________________

   ____

   ____

4. **SUMMARY:** A young girl becomes involved in the early 20th century women's suffrage movement.

   a. **GENRE:** ______________________________

   b. **MAKE A PREDICTION** (name at least one thing you would expect from this story - a character, moment, or trope) based on the summary you read and your understanding of **genre:** ______________________________

   ____

   ____

FITNESS

Please be aware of your environment and be safe at all times. If you cannot do an exercise, just try your best.

Repeat these exercises **4 ROUNDS**

2 - Lunges: 8 times to each leg.
Note: Use your body weight or books as weight to do leg lunges.

1 - Abs: 20 times

3 - Plank: 20 sec.

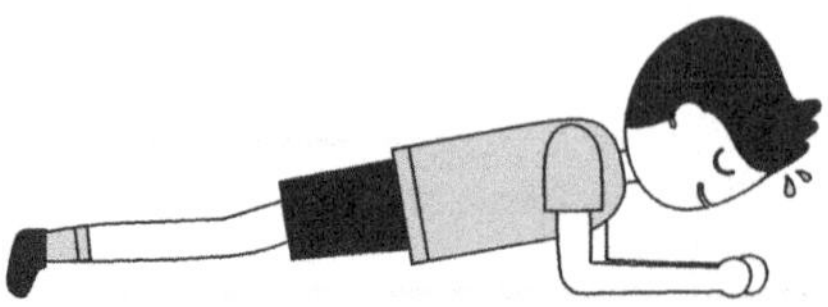

4 - Run: 50m
Note: Run **25** meters to one side and **25** meters back to the starting position.

READING PASSAGE

From "Myths and Legends of Ancient Greece and Rome"

By E.M. Berens

(Continued from Day 1's Passage)

Co-existent with Uranus and Gaea were two mighty powers who were also the offspring of Chaos. These were Erebus (Darkness) and Nyx (Night), who formed a striking contrast to the cheerful light of heaven and the bright smiles of earth. Erebus reigned in that mysterious world below where no ray of sunshine, no gleam of daylight, nor vestige of health-giving terrestrial life ever appeared. Nyx, the sister of Erebus, represented Night, and was worshipped by the ancients with the greatest solemnity.

Uranus was also supposed to have been united to Nyx, but only in his capacity as god of light, he being considered the source and fountain of all light, and their children were Eos (Aurora), the Dawn, and Hemera, the Daylight. Nyx again, on her side was also doubly united, having been married at some indefinite period to Erebus.

In addition to those children of heaven and earth already enumerated, Uranus and Gaea produced two distinctly different races of beings called Giants and Titans. The Giants personified brute strength alone, but the Titans united to their great physical power intellectual qualifications variously developed. There were three Giants, Briareus, Cottus, and Gyges, who each possessed a hundred hands and fifty heads, and were known collectively by the name of the Hecatoncheires, which signified hundred-handed. These mighty Giants could shake the universe and produce earthquakes; it is therefore evident that they represented those active subterranean forces to which allusion has been made in the opening chapter. The Titans were twelve in number; their names were: Oceanus, Ceos, Crios, Hyperion, Iapetus, Cronus, Theia, Rhea, Themis, Mnemosyne, Phoebe, and Tethys.

Now Uranus, the chaste light of heaven, the essence of all that is bright and pleasing, held in abhorrence his crude, rough, and turbulent offspring, the Giants, and moreover feared that their great power might eventually prove hurtful to himself. He therefore hurled them into Tartarus, that portion of the lower world which served as the subterranean dungeon of the gods. In order to avenge the oppression of her children, the Giants, Gaea instigated a conspiracy on the part of the Titans against Uranus, which was carried to a successful issue by her son Cronus. He wounded his father, and from the blood of the wound which fell upon the earth sprang a race of monstrous beings also called Giants.

Assisted by his brother-Titans, Cronus succeeded in dethroning his father, who, enraged at his defeat, cursed his rebellious son, and foretold to him a similar fate. Cronus now became invested with supreme power, and assigned to his brothers offices of distinction, subordinate only to himself. Subsequently, however, when, secure of his position, he no longer needed their assistance, he basely repaid their former services with treachery, made war upon his brothers and faithful allies, and, assisted by the Giants, completely defeated them, sending such as resisted his all-conquering arm down into the lowest depths of Tartarus.

WEEK 1
DAY 2

OVERVIEW OF ENGLISH CONCEPTS

READING PASSAGE

1. How were the **Giants** different from the **Titans**?

2. According to the passage, how are the characters **Uranus**, **Nyx**, and **Erebus** connected?

3. What is the main reason Uranus sends the **Giants** to Tartarus?

   A. He was scared they might try to fight him.
   B. He thought they were really ugly.
   C. He did not like having to care for his children.
   D. He wanted to punish them for their bad deeds.

4. What does the author mean when he writes, "Gaea instigated a conspiracy on the part of the Titans against Uranus, which was carried to a successful issue by her son Cronus."

   A. Cronus made a plan to defeat Uranus, and Gaea carried it out.
   B. Uranus made a plan to defeat Cronus, and Gaea carried it out.
   C. Gaea made a plan to defeat Uranus, and Cronus carried it out.
   D. Gaea made a plan to defeat Cronus, and Uranus carried it out.

5. In what ways could the character of **Cronus** be seen as a hypocrite or tyrant?

Thinking About Your Favorite Genres

Directions: Think about one of your **favorite texts of all time**. It could be a book, a movie, an episode of a TV show, or any other kind of literary text you consume! Think about your text with an emphasis on **genre** and answer the questions below to identify and explain how that genre affects the content of the text!

1. **TEXT TITLE:** ______________________________

   - Write a **short summary** (like the ones from the Day 1 activity) of the text:

     ____

     ____

     ____

     ____

   - How would you describe the **GENRE** of the text: ______________________________

     ____

   - What are some examples of **TROPES** from that genre that are present in the text?

     ____

     ____

     ____

     ____

WEEK 1
DAY 2

OVERVIEW OF ENGLISH CONCEPTS ACTIVITIES

- Why do you think the text is a **great example** of that genre?

____

____

____

____

FITNESS

Please be aware of your environment and be safe at all times. If you cannot do an exercise, just try your best.

Repeat these exercises **4 ROUNDS**

2 - Side Bending: 15 times to each side. Note: try to touch your feet.

1 - Squats: 20 times.
Note: imagine you are trying to sit on a chair.

3 - Tree Pose:
Stay as long as possible.
Note: do the same with the other leg.

MATH

Ratios and Proportional Relationships

1. A train traveled $\frac{1}{8}$ of the distance between two cities in $\frac{1}{4}$ of an hour. How long does it take the train to travel between two cities?

   A. $\frac{1}{2}$ hour
   B. 1 hour
   C. 2 hours
   D. 4 hours

2. Elizabeth got charged $89.75 for $12\frac{1}{2}$ gallons of gas. How much did she pay for one gallon of gas?

   A. $6.24
   B. $6.88
   C. $7.08
   D. $7.18

3. A runner goes 100 meters in $12\frac{1}{2}$ seconds. What is his speed in meters per second?

   A. 8 meters per sec
   B. 10 meters per sec
   C. 12 meters per sec
   D. 16 meters per sec

4. A teacher gets a bag of lollipops for her class. She buys a large bag of 84 lollipops. If she wants each child to have 3 lollipops, how many students must be in her class?

   A. 28
   B. 29
   C. 30
   D. 31

5. Joseph wants to mix red and yellow paint. He uses 11.2 mL of red paint and 5.6 mL of yellow paint. If he wants to have at least 50 mL of paint, how many batches of paint should he mix?

   A. 2
   B. 3
   C. 1
   D. 2.5

6. Emily is planning a trip between two cities. On the map, the cities are 6.5 cm apart. If the scale of the map reads 1 cm= 14 miles, how far apart are the two cities?

   A. 76 miles
   B. 84 miles
   C. 91 miles
   D. 98 miles

7. Alex pays a tip for his haircut. He pays a total of $18. If his haircut cost $15, what percentage of his total was his tip?

   A. 11%
   B. 13%
   C. 15%
   D. 17%

8. On a survey of Ocean Middle School, 36 students have only 1 sister and 45 students have only 1 brother. If 150 students were surveyed and the results are proportional of the total population of 600 students, how many students of the total population have sisters?

   A. 420
   B. 180
   C. 144
   D. 456

9. Sean gets 15 stars every week for completing his chores. For every 5 stars, he earns $1. If he wants enough money to go see a movie for $15, how many weeks does he need to work?

   A. 5
   B. 3
   C. 7
   D. 11

WEEK 1 DAY 3

MATH

Ratios and Proportional Relationships

10. The table shows a proportional relationship between x and y. What equation would represent this relationship?

x	1	3	5
y	7	21	35

A. $x=7y$
B. $y=7x$
C. $y=x+7$
D. $x=7-y$

11. The table shows a proportional relationship between x and y. What equation would represent this relationship?

x	2	8	11
y	12	48	66

A. $x=6y$
B. $y=6x$
C. $y=\frac{1}{6}x$
D. $y=\frac{x}{6}$

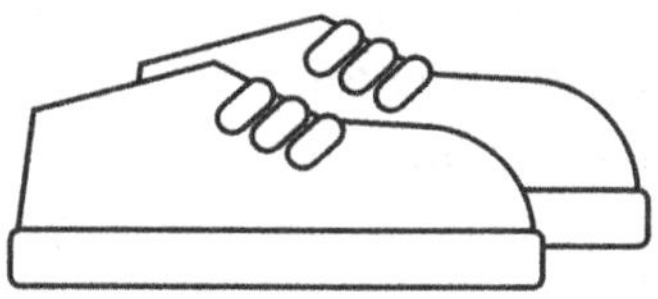

12. The table shows a proportional relationship between x and y. What equation would represent this relationship?

x	60	12	24
y	20	4	8

A. $y=\frac{1}{3}x$
B. $y=3x$
C. $y=9x$
D. $y=x+3$

13. Who earns the most per hour?

A. Paul earns $384.40 in a 40 hour work week.
B. Aaron earns $67.72 in 6 hours.
C. Alicia earns $10.23 an hour.
D. Sara earns $101.28 in 8 hours.

14. Who runs the fastest per race?

A. Connor runs 3 miles in 24 minutes.
B. Taylor runs a 6 minute mile.
C. Morgan runs 5 miles in 38.75 minutes.
D. Jayden runs three miles, three times a week, for a total workout of 58.5 minutes.

FITNESS

Please be aware of your environment and be safe at all times. If you cannot do an exercise, just try your best.

Repeat these exercises **4 ROUNDS**

1 - Bend forward: 15 times.
Note: try to touch your feet. Make sure to keep your back straight and if needed you can bend your knees.

2 - Lunges: 12 times to each leg.
Note: Use your body weight or books as weight to do leg lunges.

3 - Plank: 20 sec.

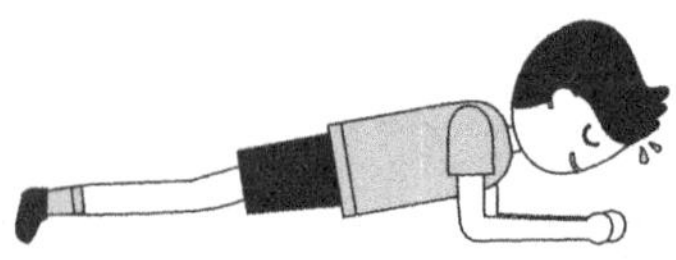

4 - Abs: 20 times

WEEK 1 DAY 4 MATH

Ratios and Proportional Relationships

1. Which equation has a constant of proportionality of 5?

A. $y=5-x$ **C.** $y=10x$
B. $y=x+5$ **D.** $y=5x$

2. What is the constant of proportionality of the equation $y=24x$?

A. 24 **C.** 4
B. 12 **D.** 2

3. Which equation has a constant of proportionality equal to $\frac{1}{4}$?

A. $y=x+\frac{1}{4}$ **C.** $y=\frac{1}{4}x$
B. $y=\frac{1}{4}-x$ **D.** $y=4x$

4. What is the constant of proportionality of the table?

x	16	24	48
y	10	15	30

A. $\frac{8}{5}$ **C.** $\frac{5}{8}$
B. $\frac{1}{8}$ **D.** $\frac{1}{5}$

5. What is the constant of proportionality of the table?

x	3	16	22
y	54	288	396

A. 16 **C.** 20
B. 18 **D.** 22

6. What is the constant of proportionality of the table?

x	14	35	77
y	6	15	33

A. $\frac{3}{7}$ **C.** $\frac{1}{7}$
B. $\frac{1}{3}$ **D.** $\frac{7}{3}$

7. Stacy planted a sunflower and kept track of how fast it grew. She used the equation $g=1.75w$, where $g=$ the height of the plant and $w=$ weeks to keep track of the growth. What does the 1.75 represent?

A. Height of the plant
B. Weeks it has grown
C. Height at the beginning
D. Amount it grows every week

8. Stacy planted a sunflower and kept track of how fast it grew. She used the equation $g=1.75w$, where $g=$ the height of the plant and $w=$ weeks to keep track of the growth. How high was the flower when it started to grow?

A. -1.75 **C.** 0
B. 3.5 **D.** 1.75

9. Stacy planted a sunflower and kept track of how fast it grew. She used the equation $g=1.75w$, where $g=$ the height of the plant and $w=$ weeks to keep track of the growth. After 9 weeks, how high has the flower grown?

A. 15.75 **C.** 11.75
B. 13.75 **D.** 8.75

WEEK 1 DAY 4 MATH

Ratios and Proportional Relationships

10. A summer camp requires 2 counselors for every 16 campers. Which equation represents that relationship?

A. $2y=16x$ **C.** $y=2x$
B. $y=16x$ **D.** $2y=8x$

11. A summer camp requires 2 counselors for every 16 campers. If there are 120 campers, how many counselors should be hired?

A. 4 **C.** 15
B. 8 **D.** 32

12. A summer camp requires 2 counselors for every 16 campers. If they have hired 12 counselors, what is the maximum number of campers they should enroll?

A. 36 **C.** 48
B. 24 **D.** 96

13. Twenty two candy bars cost $19.48. Which proportion represents how to find the cost of 15 candy bars?

A. $\frac{19.48}{15}$ **C.** $\frac{15}{22}$
B. $\frac{22}{19.48}$ **D.** $\frac{15}{19.48}$

14. 3 magazines cost $6.75. Which proportion represents how to find the cost of 8 magazines?

A. $\frac{3}{8}$ **C.** $\frac{8}{3}$
B. $\frac{6.75}{3}$ **D.** $\frac{6.75}{8}$

15. 12 folders cost $9. How much would 15 folders cost?

A. $9.75 **C.** $20
B. $10.25 **D.** $11.25

FITNESS

Please be aware of your environment and be safe at all times. If you cannot do an exercise, just try your best.

Repeat these exercises **4 ROUNDS**

1 - High Plank: 25 sec.

3 - Waist Hooping: 20 times.
Note: if you do not have a hoop, pretend you have an imaginary hoop and rotate your hips 10 times.

2 - Chair: 20 sec.
Note: sit on an imaginary chair, keep your back straight.

4 - Abs: 25 times

WEEK 1 DAY 5 MATH

Ratios and Proportional Relationships

1. Solve for x: $\frac{x}{9} = \frac{1}{3}$

   A. $x=6$
   B. $x=5$
   C. $x=4$
   D. $x=3$

2. Solve for x: $\frac{2}{12} = \frac{6}{x}$

   A. $x=6$
   B. $x=12$
   C. $x=24$
   D. $x=36$

3. Solve for x: $\frac{2}{7} = \frac{x}{35}$

   A. $x=10$
   B. $x=5$
   C. $x=15$
   D. $x=20$

4. Solve for x: $\frac{10}{15} = \frac{x}{3}$

   A. $x=1$
   B. $x=2$
   C. $x=3$
   D. $x=4$

5. Solve for x: $\frac{8}{x} = \frac{4}{7}$

   A. $x=8$
   B. $x=10$
   C. $x=14$
   D. $x=12$

6. Solve for x: $\frac{x}{6} = \frac{10}{5}$

   A. $x=60$
   B. $x=2$
   C. $x=4$
   D. $x=12$

7. Solve for x: $\frac{1}{4} = \frac{9}{x}$

   A. $x=36$
   B. $x=45$
   C. $x=54$
   D. $x=63$

8. Solve for x: $\frac{8}{11} = \frac{64}{x}$

   A. $x=80$
   B. $x=34$
   C. $x=88$
   D. $x=92$

9. Solve for x: $\frac{4}{8} = \frac{x}{4}$

   A. $x=1$
   B. $x=2$
   C. $x=3$
   D. $x=4$

10. Solve for x: $\frac{8}{x} = \frac{4}{6}$

    A. $x=6$
    B. $x=8$
    C. $x=10$
    D. $x=12$

WEEK 1 DAY 5

MATH

Ratios and Proportional Relationships

11. Solve for x: $\frac{x}{25} = \frac{2}{5}$

A. x=10
B. x=8
C. x=5
D. x=2

12. Solve for x: $\frac{27}{x} = \frac{9}{4}$

A. x=8
B. x=9
C. x=12
D. x=15

13. Solve for x: $\frac{9}{1} = \frac{9}{x}$

A. x=1
B. x=9
C. x=3
D. x=2

14. Solve for x: $\frac{3}{x} = \frac{6}{18}$

A. x=3
B. x=9
C. x=6
D. x=18

15. Solve for x: $\frac{40}{10} = \frac{x}{2}$

A. x=2
B. x=6
C. x=4
D. x=8

YOGA

Please be aware of your environment and be safe at all times. If you cannot do an exercise, just try your best.

1 - Down Dog: 30 sec.

2 - Bend Down: 30 sec.

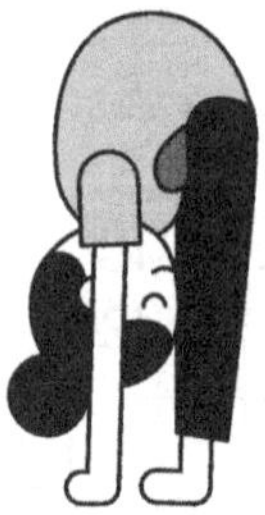

3 - Chair: 30 sec.

4 - Child Pose: 30 sec.

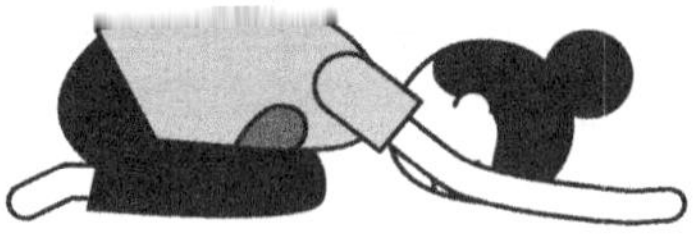

5 - Shavasana: as long as you can. Note: think of happy moments and relax your mind.

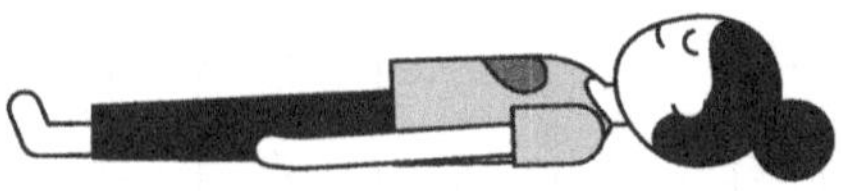

WEEK 1 DAY 6

EXPERIMENT

Newton's First Law

*In our upcoming science activities, we'll be thinking about some of the foundational principles of **physics**. **Physics** is the study of matter, energy, and motion. It focuses on how and why objects move the way they do. One of the most important figures in the study of physics is **Isaac Newton**, who lived from 1643-1727. Newton developed **Three Laws of Motion** that help us understand how, why, and when things move.*

*This week, we'll focus on Newton's first law, which states: **An object at rest will stay at rest unless acted upon by an outside force. An object in motion will stay in motion at a constant speed until acted upon by an outside force.** This sounds like a complex concept, but we can access it using a very easy demonstration/experiment!*

Materials:

- 1 hardcover book (something like a dictionary or encyclopedia - the bigger the better!)
- A desk or table
- Notepaper

Procedure:

1. Find a desk or table with a flat top and a large book. Look at both, and write down a quick prediction on your notepaper of what will happen when you place the heavy book on the desk. Try to describe what kind of **motion** you expect will occur.

2. Pick up your heavy book and gently place it on the table. Observe what happens and write the results down in your notes. When the object sits on the table, does it **move**? As long as the table is flat and nobody pushes, pulls, or tips anything, the book should sit perfectly still, demonstrating that **an object at rest will stay at rest unless acted upon by an outside force.**

3. Next, on your notepaper, write down a short prediction explaining what you think will happen if you pick the book up off the table and drop it. What kind of **motion** do you think will occur?

4. Pick your heavy book up off the table and carry it a few feet away so that you're standing over the floor (ideally a carpeted floor to reduce noise). Gently release it with your hands, allowing it to fall downward. Observe the **motion** of the book and how that motion **stops**. Write a brief description of what you witnessed in your notes. When you let go of the book, **an outside force (gravity) pulled it downward at a constant speed, until an outside force (provided by your floor) brought it back to rest,** where it will remain until you pick it back up.

5. Answer the questions below to extend your thinking about Newton's First Law and be sure to pick your heavy book up off the floor!

WEEK 1 DAY 6 EXPERIMENT

Follow-Up Questions:

1. How can you tell when an object is "**at rest**"?

2. What are some other **forces** you can think of (besides the example of gravity we used) that might cause an object to move?

YOGA

Please be aware of your environment and be safe at all times. If you cannot do an exercise, just try your best.

2 - Down Dog: 30 sec.

3 - Stretching: Stay as long as possible.
Note: do on one leg then on another.

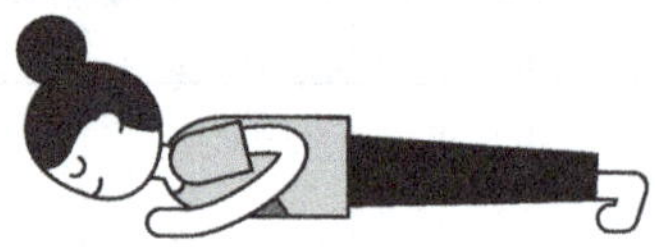

4 - Lower Plank: 20 sec.
Note: Keep your back straight and body tight.

1 - Tree Pose: Stay as long as possible.
Note: do on one leg then on another.

6 - Shavasana: 5 min.
Note: this pose is very important and provides you with long term benefits. Try not to skip this. Close your eyes and imagine who you want to be and what your goals are! Always think happy thoughts.

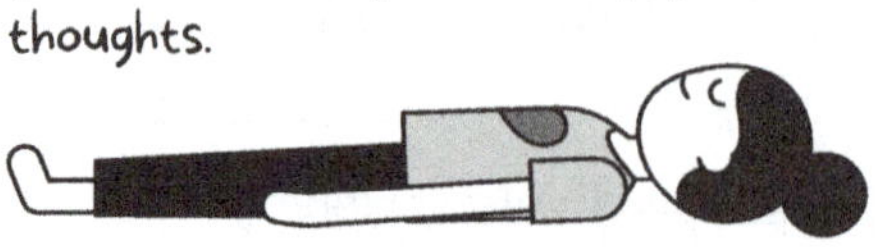

5 -Book Pose: 20 sec.
Note: Keep your core tight. Legs should be across from your eyes.

LOGIC

TASK:

Using the stairs and ladders, find the way from point 1 to point 2.
Do not jump or climb up!

WEEK 2

ARGOPREP

WEEK 2 DAY 1 OVERVIEW OF ENGLISH CONCEPTS

Explicit vs. Implicit Meaning

As you've matured as a reader, you've graduated from reading very short, simple, and straightforward texts to longer, more complex pieces. Along the way, you've seen **figurative language**, like metaphors and similes, introduced to create meaning in new ways without just saying things directly. Now it's time to think even deeper and look at ways that authors construct meaning far beyond what the words in the text say.

Most texts - at least most good texts - have an explicit meaning. In **informational and non-fiction texts**, there is often a section at the very beginning that introduces the topic and explains what aspects of that topic the author will explore. That's an example of being very **explicit**! In **literary texts and stories**, the **explicit** level of the text includes all the simple facts of plot comprehension and what the characters say and do.

Many texts, however, also contain **implicit meaning** - ideas that are suggested, implied, or talked "around" by the author. Sometimes, authors use **implicit** meanings to avoid saying things too directly or to create a fun "nugget" for close readers who understand what they mean. It's really important to be able to notice and identify **implicit** ideas in a text to make sure that you actually understand what's going on!

Key Terms:

Explicit Meaning: The clearly-stated goal, purpose, or message of the text
Implicit Meaning: The suggested or implied goal, purpose, or message of the text which is not explicitly stated.
Inference: A conclusion drawn based on evidence
Inferential Thinking: Using evidence to draw conclusions about implicit meaning.

OVERVIEW OF ENGLISH CONCEPTS

Using Inferential Thinking to Access Implicit Meaning:

- Finding **explicit meaning** should be relatively straightforward
    - o Focus on the **words** of the text. Use references to define any you're unfamiliar with, and use reading comprehension strategies you've built over the years to understand the **literal meanings of the words**
- Once you've got the **explicit meaning** out of the way, it's time to dive in and consider the **implicit meaning**
    - o Study the **explicit meaning** and ask yourself: Are there any logical "next steps" or "missing steps" in the author's thinking that they are omitting or leaving out? Why might they be leaving that out?
    - o Ask yourself: Does the author's wording or thinking suggest they have certain beliefs or motives that they **are not** saying outright?
    - o Are there any examples of **wordplay** or places where things could be **interpreted two or more different ways?** (We'll refine these skills even more in **Week 3!**)
- One good way to access implicit meaning is to **talk about texts** with your friends, family members, or other people you know are reading the same things.
    - o Did everybody get the same things out of the text? Did one reader notice something that you missed? Were there spots where people interpreted or understood the text in different ways? Did other people make **inferences** about things that seemed to be suggested that you may not have noticed?

READING PASSAGE

From "Myths and Legends of Ancient Greece and Rome"

By E.M. Berens

(Continued from Week 1's Passages)

Cronus was the god of time in its sense of eternal duration. He married Rhea, daughter of Uranus and Gaea, a very important divinity, to whom a special chapter will be devoted hereafter. Their children were, three sons: Aides (Pluto), Poseidon (Neptune), Zeus (Jupiter), and three daughters: Hestia (Vesta), Demeter (Ceres), and Hera (Juno). Cronus, having an uneasy conscience, was afraid that his children might one day rise up against his authority, and thus verify the prediction of his father Uranus. In order, therefore, to render the prophecy impossible of fulfilment, Cronus swallowed each child as soon as it was born, greatly to the sorrow and indignation of his wife Rhea. When it came to Zeus, the sixth and last, Rhea resolved to try and save this one child at least, to love and cherish, and appealed to her parents, Uranus and Gaea, for counsel and assistance. By their advice she wrapped a stone in baby-clothes, and Cronus, in eager haste, swallowed it, without noticing the deception. The child thus saved, eventually, as we shall see, dethroned his father Cronus, became supreme god in his stead, and was universally venerated as the great national god of the Greeks.

Anxious to preserve the secret of his existence from Cronus, Rhea sent the infant Zeus secretly to Crete, where he was nourished, protected, and educated. A sacred goat, called Amalthea, supplied the place of his mother, by providing him with milk; nymphs, called Melissae, fed him with honey, and eagles and doves brought him nectar and ambrosia. He was kept concealed in a cave in the heart of Mount Ida, and the Curetes, or priests of Rhea, by beating their shields together, kept up a constant noise at the entrance, which drowned the cries of the child and frightened away all intruders. Under the watchful care of the Nymphs the infant Zeus throve rapidly, developing great physical powers, combined with extraordinary wisdom and intelligence. Grown to manhood, he determined to compel his father to restore his brothers and sisters to the light of day, and is said to have been assisted in this difficult task by the goddess Metis, who artfully persuaded Cronus to drink a potion, which caused him to give back the children he had swallowed. The stone which had counterfeited Zeus was placed at Delphi, where it was long exhibited as a sacred relic.

1. **Why** do you think Cronus was "afraid his children might one day rise up against his authority"?

2. How did **Cronus** attempt to prevent his children from betraying and defeating him?

____

____

3. According to the text, who was "venerated as the great national god of the Greeks"?

   A. Cronus
   B. Uranus
   C. Rhea
   D. Zeus

4. What was **Metis'** role in Rhea and Zeus' plan?

   A. Metis nurses Zeus instead of his mother while he is hiding on Crete.
   B. Metis poisons Cronus, which makes him throw up his children.
   C. Metis made noise to prevent Cronus from hearing baby Zeus' cries.
   D. Metis helps Zeus train to become stronger.

5. Which of Zeus' traits described in this passage predict or foreshadow that he will be a **great leader** and an **important ruler**?

____

____

____

____

____

____

Thinking About Implicit Meaning (Pt. 1)

Directions: Read each short paragraph below and consider the **implicit meaning** of the situation that's taking place. After you've read, answer the questions below to demonstrate your understanding of what's really going on in the text!

Alyssa was sitting alone by herself in an empty coffee shop. She was quietly sipping tea and working on her laptop when another customer came in. After ordering his coffee, the new customer sat right next to her, even though there were lots of other seats, and began setting up his own laptop. As he reached down below Alyssa's legs to plug in his power cord, Alyssa said, "I think there's a second outlet down by that other table."

1. Which of these is the **explicit** meaning of Alyssa's words?

A. It might be more convenient for you to use that other empty plug.
B. You should go to another coffee shop.
C. There's no need for us to be so close together, given that the coffee shop is empty.
D. I don't want to overload the outlet by plugging two laptops into it.

2. Which of these is the **implicit** meaning of Alyssa's words?

A. It might be more convenient for you to use that other empty plug.
B. You should go to another coffee shop.
C. There's no need for us to be so close together, given that the coffee shop is empty.
D. I don't want to overload the outlet by plugging two laptops into it.

Last week, my mom was making her famous leek soup. She sent my dad and I to the grocery store because she needed a few more leeks. When we went to the produce section, they only had three left, but we bought them all. While we were walking back to our car in the parking lot, I dropped the bag, and all the leeks fell out onto the ground. "Oh no! Now your mom is going to put us in the soup!" my dad exclaimed.

ACTIVITIES

3. Which of these is the **explicit** meaning of the father's words?

- **A.** Now your mother is going to cook us instead of the leeks!
- **B.** Now we are in big trouble!
- **C.** Now your mother is going to replace the leeks in the recipe with something random, which will result in a strange, different soup.
- **D.** The father's words don't really have an explicit meaning, as they are entirely nonliteral.

4. Which of these is the **implicit** meaning of the father's words?

- **A.** Now your mother is going to cook us instead of the leeks!
- **B.** Now we are in big trouble!
- **C.** Now your mother is going to replace the leeks in the recipe with something random, which will result in a strange, different soup.
- **D.** The father's words don't really have an implicit meaning, as they are entirely literal.

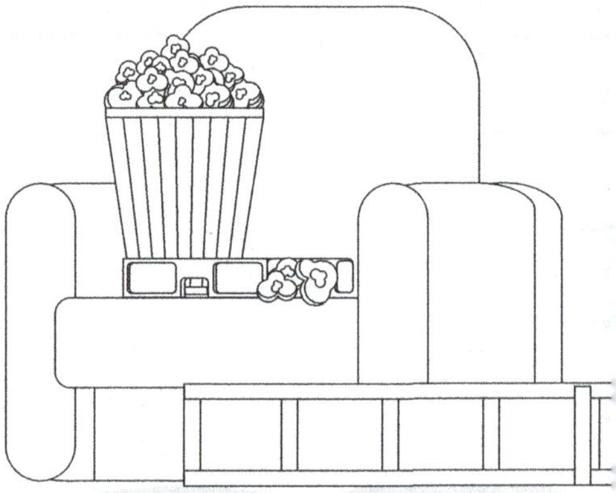

FITNESS

Please be aware of your environment and be safe at all times. If you cannot do an exercise, just try your best.

Repeat these exercises **4 ROUNDS**

1 - Abs: 20 times

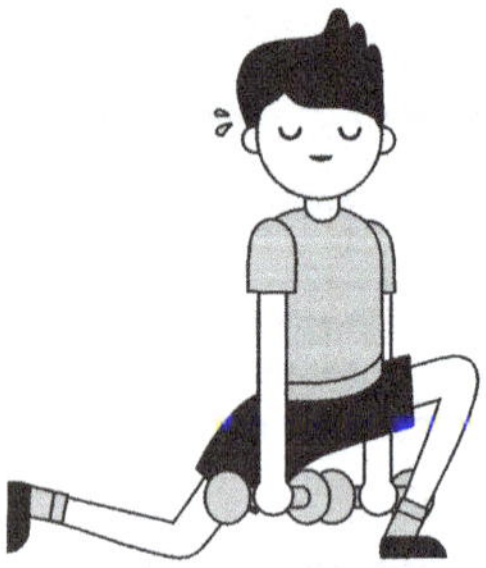

2 - Lunges: 8 times to each leg.
Note: Use your body weight or books as weight to do leg lunges.

3 - Plank: 20 sec.

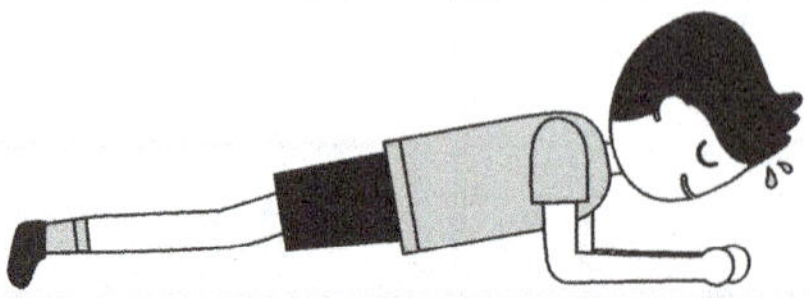

4 - Run: 50m
Note: Run 25 meters to one side and 25 meters back to the starting position.

READING PASSAGE

From Myths and Legends of Ancient Greece and Rome

By E.M. Berens

(Continued from Day 1's Passage)

Cronus was so enraged at being circumvented that war between the father and son became inevitable. The rival forces ranged themselves on two separate high mountains in Thessaly; Zeus, with his brothers and sisters, took his stand on Mount Olympus, where he was joined by Oceanus, and others of the Titans, who had forsaken Cronus on account of his oppressions. Cronus and his brother-Titans took possession of Mount Othrys, and prepared for battle. The struggle was long and fierce, and at length Zeus, finding that he was no nearer victory than before, bethought himself of the existence of the imprisoned Giants, and knowing that they would be able to render him most powerful assistance, he hastened to liberate them. He also called to his aid the Cyclops (sons of Poseidon and Amphitrite), who had only one eye each in the middle of their foreheads, and were called Brontes (Thunder), Steropes (Lightning), and Pyracmon (Fire-anvil). They promptly responded to his summons for help, and brought with them tremendous thunderbolts which the Hecatoncheires, with their hundred hands, hurled down upon the enemy, at the same time raising mighty earthquakes, which swallowed up and destroyed all who opposed them.

Aided by these new and powerful allies, Zeus now made a furious onslaught on his enemies, and so tremendous was the encounter that all nature is said to have throbbed in accord with this mighty effort of the celestial deities. The sea rose mountains high, and its angry billows hissed and foamed; the earth shook to its foundations, the heavens sent forth rolling thunder, and flash after flash of death-bringing lightning, whilst a blinding mist enveloped Cronus and his allies.

And now the fortunes of war began to turn, and victory smiled on Zeus. Cronus and his army were completely overthrown, his brothers dispatched to the gloomy depths of the lower world, and Cronus himself was banished from his kingdom and deprived forever of the supreme power, which now became vested in his son Zeus. This war was called the Titanomachia, and is most graphically described by the old classic poets.

1. Based on this passage, and the passages from last week, **why** do you think some of the Titans sided with Zeus against their brother Cronus?

2. How was the war going for Zeus **before he released the Giants from Tartarus**?

3. Which of these statements is **not** true about Brontes, Steropes, and Pyracmon?

A. They are children of Poseidon.
B. They each only have one eye.
C. They are among the Titans.
D. They fight using thunderbolts.

4. Based on the last sentence of the passage, which of these statements is definitely true?

A. More than one author has written about this story.
B. The old classic poets wrote the best version of the story told in this passage.
C. The version of the story told in this passage is much more detailed than the ones that were told by the old classic poets.
D. The old classic poets were the greatest storytellers of the ancient world.

5. Using your own words, describe what the battle between Zeus and Cronus' armies **looked like**:

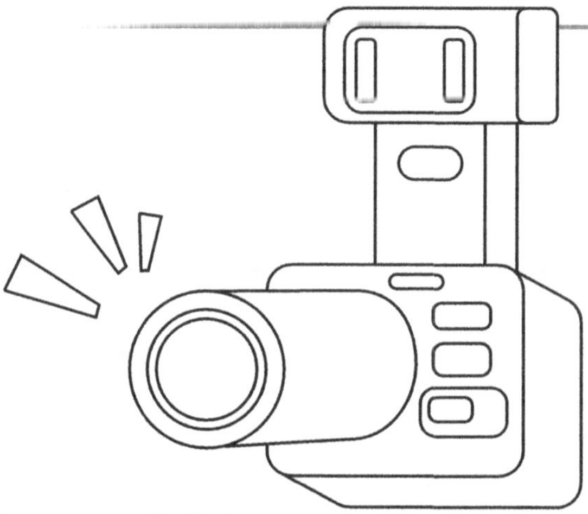

Thinking About Implicit Meaning (Pt. 2)

Directions: Read each short paragraph below and consider what is being said or described both **explicitly** and **implicitly**. After you've read and thought about each paragraph, answer the questions below to explain its meaning.

1. My Auntie Liz isn't really my mother's sister, but I have addressed her by that name my whole life. She and my mom first met in college, and they have been friends ever since. Auntie Liz is a very respected art restoration expert and works at one of the country's biggest and most famous museums. Because she lives in the world of art and intellectuals, Auntie Liz can be a little hard to communicate with. She knows a lot about art but seems to have some difficulty identifying with other people's situations. Anytime she and my mother hang out, they wind up staying up all night talking. It's the only time my mom seems to let loose.

   Suppose you knew the author of this paragraph **didn't actually like** Auntie Liz very much. How does the paragraph **implicitly** suggest that idea without ever stating it? What hints does the author provide the reader? Try to name at least **3!**

   a. ______________________________

   ____

   b. ______________________________

   ____

   c. ______________________________

   ____

2. We have two dogs: Reilly and Snarf. Reilly is a mixed breed male dog, who we have had for five years, and Snarf is a female dachshund we have had for three years. Snarf was originally named "Flower," but we changed her name because of the characteristic whining noise she makes all the time. The two of them are best friends, but they also fight all the time. Reilly is super energetic and always wants to play, but Snarf likes to nap and relax more. That means sometimes when Reilly sneaks up on her, Snarf will snap at him because she wants to be left alone. Outdoors, Reilly loves to play with us and chase a frisbee or ball, but Snarf is always sniffing around, trying to hunt mice or squirrels. I'm always a little nervous about what will happen if she catches one! Regardless of their differences, we love Reilly and Snarf equally, and I consider both of them the best dogs ever.

OVERVIEW OF ENGLISH CONCEPTS
ACTIVITIES

In the final sentence, the author claims they love and enjoy Reilly and Snarf equally and consider them both "the best dogs ever." What **implicit** ideas from the text challenge that claim? Try to cite at least **3** hints from the text that shows the author feels differently about Reilly and Snarf:

a. ______________________________

____

b. ______________________________

____

c. ______________________________

____

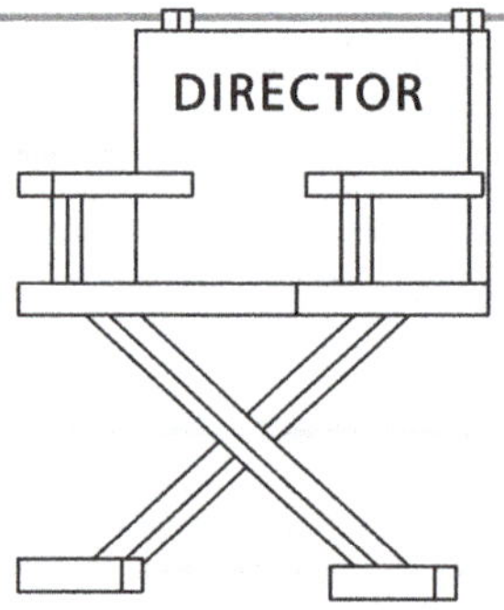

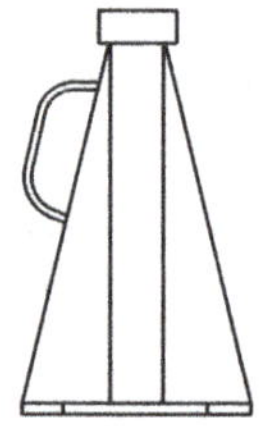

FITNESS

Please be aware of your environment and be safe at all times. If you cannot do an exercise, just try your best.

Repeat these exercises **4 ROUNDS**

1 - Squats: 20 times.
Note: imagine you are trying to sit on a chair.

2 - Side Bending: 15 times to each side. Note: try to touch your feet.

3 - Tree Pose: Stay as long as possible.
Note: do the same with the other leg.

WEEK 2 DAY 3 MATH

Word Problems

1. Emma bought a pair of headphones that were originally $65. If they were discounted 40%, how much was the new price?

   A. $14 C. $26
   B. $22 D. $39

2. Ethan wanted to buy 2 pairs of jeans that were discounted 20% off the original price. If the original price of a pair of jeans was $89, how much would he spend on the two pair?

   A. $35.60 C. $71.20
   B. $142.40 D. $17.80

3. Greg had $45 to buy a video game. There was no tax, and the game was originally priced at $65. He had a 20% off coupon to use. How much more does he need to buy the game?

   A. $7 C. $9
   B. $8 D. $10

4. Jessie wants a new computer. The original price was $399. If it is discounted 15%, what is the new price of the computer?

   A. $339.15 C. $59.85
   B. $229.15 D. $85.59

5. John bought a new hat. It was originally $52 but was sold for $47.84 (not including tax). What percent discount did John receive?

   A. 4% C. 8%
   B. 6% D. 10%

6. Sherry bought tickets to a baseball game. She bought 5 tickets and received a 10% discount on them all. If the original ticket price was $84, what was the total price for all 5 tickets?

   A. $37 C. $420
   B. $42 D. $378

7. A bookstore received a shipment of 24 books for $96. If they are going to sell the books for $6.50, what is the markup percentage?

   A. 45% C. 78%
   B. 63% D. 95%

8. A farmer sells a bundle of 80 apples to a store for $36. If the store wants to make a profit of $0.20 an apple, how much will each apple be priced for?

   A. 65 cents C. 55 cents
   B. 60 cents D. 50 cents

9. A t-shirt factory sells t-shirts for $9.00 to clothing stores. If the clothing store marks the shirts up by 15%, how much do customers pay for each shirt?

   A. $9.35 C. $10.15
   B. $9.85 D. $10.35

10. Our school store buys a shipment of 1000 pencils for $65. If we sell each pencil for 10 cents, what is the markup percentage of each pencil?

    A. 28% C. 54%
    B. 43% D. 65%

WEEK 2 DAY 3

MATH

Word Problems

11. A factory sells pasta boxes to a supermarket for $1.10 a box. If the supermarket marks the pasta up 15%, how much do customers pay for each pasta box?

A. $0.96 **C.** $1.00
B. $1.27 **D.** $1.25

12. At the high school basketball game, the concession stand is selling candy bars for $0.75 each. If they marked up the bars 20% and they buy the bars in boxes of 20, how much do they pay for each bar?

A. $0.45 **C.** $0.88
B. $0.63 **D.** $0.95

13. Last year, our town received 17 inches of snow. This year, we received 5 inches more. What is the percent increase of snow we received this year?

A. 5% **C.** 22.7%
B. 18.6% **D.** 29.4%

14. The track team is training for their first meet. They ran 3 miles for a total of 24 minutes at the beginning of the training period. At the end of the training period, they ran 5 miles for a total of 27 minutes. What is the percent increase of the distance they ran?

A. 67% **C.** 88%
B. 13% **D.** 29%

15. Melissa has held a yard sale for the past three years. This year, she made $276 in sales. Her first year, she made $240 in sales. What was the percent increase in her sales from the first year?

A. 3% **C.** 10%
B. 5% **D.** 15%

FITNESS

Please be aware of your environment and be safe at all times. If you cannot do an exercise, just try your best.

Repeat these exercises **4 ROUNDS**

1 - Bend forward: 15 times.
Note: try to touch your feet. Make sure to keep your back straight and if needed you can bend your knees.

2 - Lunges: 12 times to each leg.
Note: Use your body weight or books as weight to do leg lunges.

3 - Plank: 20 sec.

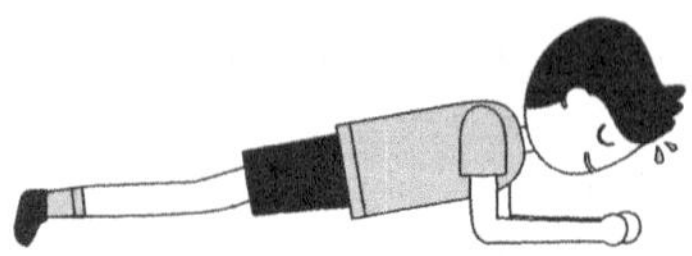

4 - Abs: 20 times

WEEK 2 DAY 4 MATH

Word Problems

1. A real estate agent earns a 3% commission on every house she sells. Last month, she sold two houses, one for $115,000 and one for $85,000. What was her total commission?

   A. $5000 C. $3450
   B. $6000 D. $2550

2. A real estate agent has a goal to earn $4,000 in a month. Her commission is 3.5% on every house she sells. How much docs she need to sell to meet her sales goal?

   A. $114,000 C. $116,000
   B. $115,000 D. $117,000

3. A salesman earns a 6% commission on furniture sold. If he sells $2500 in a week, what was his total earnings?

   A. $150 C. $1500
   B. $200 D. $2000

4. A fashion house employs sales people who earn commission. If a salesman had $875 worth of sales and earned $131.25, how much commission does he make?

   A. 9% C. 15%
   B. 12% D. 18%

5. A car salesman earns a base salary of $15 an hour plus 1% of car sales. If he works 4 hours making a sale and sells a car for $45,000, what is his total take home pay for that day?

   A. $60 B. $325
   C. $450 D. $510

6. A couple buys $500,000 in life insurance. They have to pay the insurance agency a brokerage fee of 0.7%. How much will they pay to the the agency?

   A. $350,000 C. $35,000
   B. $350 D. $3500

7. Marcus is buying a new video game. He pays d dollars for the game and then needs to pay an additional 6% tax. Which expression represents the total amount Marcus will pay for the game?

   A. 1.06d C. 0.06d
   B. 0.6d D. d+0.6d

8. Malcolm is getting food for his cat. The food costs $6.45. If he has to have at least $6.77, what is the percentage of sales tax he has to pay?

   A. 2% C. 5%
   B. 4% D. 8%

9. April is saving money for a new backpack. The one she wants costs $65. She also has to pay a tax of 8%. What is the total amount of money she needs to save?

   A. $68.41 C. $75.36
   B. $64.24 D. $70.20

10. Lauren wants to take a trip with her family. She buys a set of new suitcases for $424. If she ends up paying around $441 dollars, what percentage of sales tax was she charged?

    A. 2% C. 6%
    B. 4% D. 8%

WEEK 2 DAY 4 MATH

Word Problems

11. Five people went out to lunch. Each person's bill was around **$12**. If they want to leave a 15% tip, how much should each person leave as a tip?

A. $1.80 **C.** $1.40
B. $13.80 **D.** $12.00

12. The hairdresser charged Angela **$72** for her hairstyling. Angela wants to leave a 10% tip. Which expression represents how much she should leave for a tip?

A. 72 x 0.1 **C.** 72 + 0.1
B. 72 x 0.01 **D.** 72 x 1.1

13. Melissa picks up a carryout order that costs **$32**. If it is standard to tip 10%, how much should she leave for a tip?

A. $35.20 **C.** $0.32
B. $34.40 **D.** $3.20

14. Jackson takes a taxi trip that ends up costing **$35**. If he ends up paying **$41**, what percent did he tip his driver?

A. 15% **C.** 17%
B. 16% **D.** 18%

15. Graham and his sister went out to dinner with their family. The food total was **$86**. If that did not include the tax of 6% and a 15% tip, how much would their total cost for dinner be?

A. $104.06 **C.** $98.90
B. $91.16 **D.** $102.60

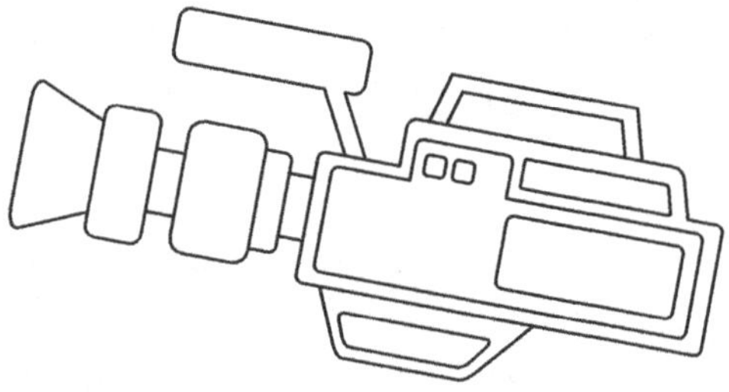

FITNESS

Please be aware of your environment and be safe at all times. If you cannot do an exercise, just try your best.

Repeat these exercises **4 ROUNDS**

2 - Chair: 20 sec.
Note: sit on an imaginary chair, keep your back straight.

4 - Abs: 25 times

3 - Waist Hooping: 20 times.
Note: if you do not have a hoop, pretend you have an imaginary hoop and rotate your hips 10 times.

1 - High Plank: 25 sec.

MATH

Word Problems

1. Last year there were 218 students in 7th grade. This year, there are 9% more. How many students are in 7th grade this year?

   A. 238 C. 1526
   B. 19 D. 234

2. Last week, Marco swam 65 laps. This week, he swam 10% less. How many fewer laps did he swim this week?

   A. 5 C. 9.5
   B. 6.5 D. 11

3. On our road trip, we drove 25% more miles in the afternoon than in the morning. If we drove 240 miles in the morning, how many miles did we drive in the afternoon?

   A. 25 C. 300
   B. 360 D. 60

4. A book is reprinted with a larger font. The old book was 242 pages. The new book is 286 pages. What was the percent increase of pages in the book?

   A. 6% C. 18%
   B. 11% D. 25%

5. Richard wants to weed 33% of his garden today. The total garden is 144 square yards. How many square yards should Richard weed to reach his goal?

   A. 28 C. 56
   B. 48 D. 96

6. Eliza walks 2 miles to school in 30 minutes. If she keeps the same pace during a race and it takes her 1 hour 15 minutes to finish, how far is the race?

   A. 5 miles C. 3 miles
   B. 4 miles D. 2 miles

7. The Business Club wants to sell lemonade at lunch on hot days. They have a recipe that calls for 10 lemons for every gallon of lemonade. The store sells lemons at a cost of 6 for $3. How much will it cost to get lemons to make 12 gallons of lemonade?

   A. $15 C. $45
   B. $30 D. $60

8. A solar panel charges a battery to last 6 hours of constant light with a 1 hour charge. If the Jackson family is camping and needs light for 12 hours, how long should they charge the panel?

   A. 120 minutes
   B. 90 minutes
   C. 60 minutes
   D. 30 minutes

9. A delivery truck goes 487.5 miles in 6.5 hours. If he is driving the same rate, what is his average speed?

   A. 70 mph
   B. 75 mph
   C. 60 mph
   D. 65 mph

WEEK 2 DAY 5 MATH

Word Problems

10. A fence is 3.5 feet tall. It needs to be another 15% tall for the yard, how tall does the fence need to be?

A. 3.6 feet
B. 3.7 feet
C. 3.8 feet
D. 4.03 feet

11. Morgan is buying a video game for $65. If the game is on sale for 20% off, which expression represents how much Morgan will pay?

A. 65 - 20
B. 52 + 20
C. 65 x 0.8
D. 65 x 0.2

12. It costs $0.24 cents to make a cup of lemonade. If Alex wants to make 25% profit, which expression represents how much he should sell the lemonade per cup?

A. 1.25 x 0.24
B. $0.30
C. 0.30 x 0.25
D. 0.24 + 0.25

13. Six ounces of orange juice contains 75 calories. How much orange juice contains 100 calories?

A. 14 ounces
B. 12 ounces
C. 10 ounces
D. 8 ounces

14. A lion's heart beats 12 times in 16 seconds. How much does a lion's heart beat in 1 minute?

A. 12
B. 45
C. 30
D. 42

15. On Tuesday, Paul had 3 times as much money as his little brother. On Wednesday, Paul gave half of his money to his brother. On Thursday, Paul had $10. How much money did Paul have on Tuesday?

A. $20
B. $30
C. $40
D. $50

YOGA

Please be aware of your environment and be safe at all times. If you cannot do an exercise, just try your best.

1 - Down Dog: 30 sec.

2 - Bend Down: 30 sec.

3 - Chair: 30 sec.

4 - Child Pose: 30 sec.

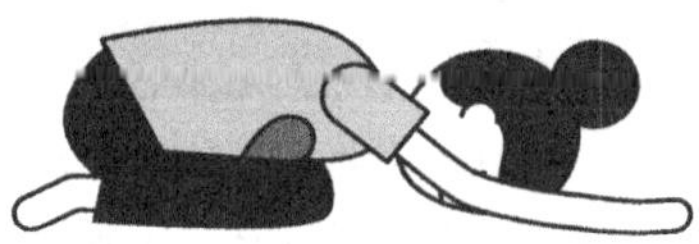

5 - Shavasana: as long as you can. Note: think of happy moments and relax your mind.

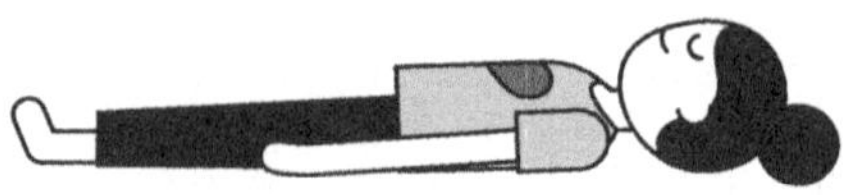

EXPERIMENT

Newton's Second Law

Last week, we introduced Newton's First Law: ***that objects at rest will stay at rest until acted upon by outside forces, while objects in motion will stay in motion, until acted upon by an outside force****. Today, we're going to look at what exactly Newton means when he's talking about* ***force****, which is defined using* ***Newton's Second Law****!*

Newton's Second Law states that ***Mass x Acceleration = Force*** *(often shortened to F = ma). Basically, that means that anything made of matter (so, anything) moving with some speed (acceleration) can create a force. For example, when a car hits something, the force of the impact is equal to the car's mass multiplied by the acceleration (speed) it had when it hit that object.*

We hinted at one of the easiest ways to understand F = ma last week, and that's by studying the force of ***gravity****!*

Materials:

- A digital scale that can be configured to measure in **grams** or **kilograms**
- A variety of small, non-breakable household objects (throw pillows, pens, dice, coins, small books, kitchen utensils, etc.)
- A medium-sized cardboard box (at least big enough to fit your largest item from above)
- 1-3 soft pillows
- A calculator
- A table

WEEK 2 DAY 6 EXPERIMENT

Procedure:

1. Use your soft pillows to line your cardboard box. You are creating a safe landing place that will protect your household objects (and your floor) when you do a drop test in a few minutes.

2. Gather your small household objects and, examining them closely, think about which ones probably require the most **gravitational energy** for the Earth to attract. On your notepaper, create a **ranking**, from most to least, of which objects you feel receive the most **force of gravity**.

3. One at a time, take your objects and **drop them** (don't throw them or knock them) from the height of the edge of your table down into the pillow-lined box below. As you do this, feel free to revisit and change your ranking as to which items you think are exhibiting the greatest gravitational pull. Think about what other clues tell you there is a lot of force being generated (sounds, etc.).

4. Gather your small household objects and place them on your digital scale one at a time, measuring their mass in **grams** or **kilograms**. Make a note of each object's weight in **kilograms** on your notepaper. If your scale can only measure in grams, **take that number and divide it by 1,000 on your calculator** to find the mass in kilograms.

5. Next, grab your calculator and a second piece of notepaper to calculate how much gravitational **force** the earth was exerting on each object during your drop test. Remember, **F = ma**. That means you need to multiply the **mass** of each object (in kilograms) by the **acceleration it was falling at.** Luckily, the acceleration of gravity is **constant**, so for each object, just multiply its mass in kilograms times 9.8 meters per second squared. Your answer will be the amount of force the Earth was exerting on your object, expressed in "Newtons" (named after Sir Isaac).

    a. For example, if you had an object that weighed 2 kilograms, you would multiple 2 x 9.8 to get 19.6 Newtons of gravitational force.

6. After you've done the math and determined the force of gravity acting on all your objects, revisit your predictions and double-check to see whether you were correct about the gravitational force acting on the objects and, if not, in which ways you were off-track. Once you've done that, clean up your materials and answer the questions below.

EXPERIMENT

Follow-Up Questions:

1. How did the **mass** of the objects connect to the **force** of gravity?

2. Which objects **surprised** you during your investigation? Were you expecting them to be experiencing more or less of a gravitational pull?

YOGA

Please be aware of your environment and be safe at all times. If you cannot do an exercise, just try your best.

1 - Tree Pose: Stay as long as possible. Note: do on one leg then on another.

2 - Down Dog: 30 sec.

3 - Stretching: Stay as long as possible. Note: do on one leg then on another.

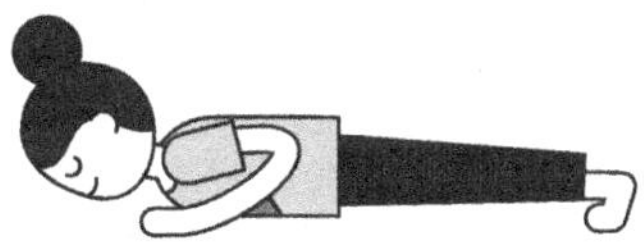

4 - Lower Plank: 20 sec. Note: Keep your back straight and body tight.

5 -Book Pose: 20 sec. Note: Keep your core tight. Legs should be across from your eyes.

6 - Shavasana: 5 min. Note: this pose is very important and provides you with long term benefits. Try not to skip this. Close your eyes and imagine who you want to be and what your goals are! Always think happy thoughts.

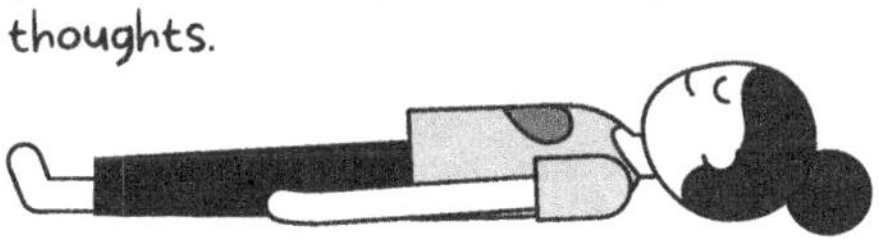

WEEK 2 DAY 7 MAZE

TASK: Color in the correct path from "Start" to "Finish".

WEEK 3

Practice makes perfect. Studying 20 minutes on a daily basis is an incredible way to strengthen your foundations. While studying, refrain from any distractions.

OVERVIEW OF ENGLISH CONCEPTS

Puns & Words with Multiple Meanings

Last week, we introduced the concept of **implicit meaning** and discussed how there's often more going on in a text than simply the straightforward **explicit** meanings of the words. This week, we'll talk specifically about how authors use words with multiple definitions or similar-sounding words to build implicit or second meanings.

If you like to make or read jokes, you probably already know some of the foundational skills of wordplay. Many punchlines involve using **homonyms**, **homophones**, **or homographs** (see below for definitions) to create a humorous effect. These jokes are easy **to hear** (especially because of the way that people tend to deliver them), but recognizing this kind of wordplay can be much trickier **as a reader** because you don't have your ear available to you as a tool. Below, we'll point out some things to notice and strategies to help you recognize puns and wordplay when you read.

Key Terms:

Homonyms: Words that sound alike but have different meanings
Homophone: Words that sound alike, have different meanings, and are spelled differently

- **NOTE:** All **homophones are homonyms**, but not all homonyms are homophones!

Homograph: Words that are spelled the same but have different meanings
Pun: A joke based on a word having two possible meanings, or a homophone/homograph that creates a completely different visual or effect than the phrase's explicit meaning.

- For example, "Two peanuts were walking down the street, and one was assaulted" is an example of a **pun** because the word "assaulted" (meaning beaten) sounds very close to "a salted," as in a salted peanut.

Identifying Puns and Wordplay as a Reader:

Like we said before, puns and wordplay are easy to appreciate with your ear, but trickier to observe with your eyes as you read. Here are a few strategies to improve your ability to notice word play being used to build **implicit** meaning:

- **Don't be afraid** to look words up as you read!
  - One of the biggest mistakes students make is assuming they're "pretty sure" what a word means and moving on. In the case of a word with multiple meanings, you might be missing out on big insight by not taking a second to look the word up online or in a dictionary.
- If you get the feeling you're missing something clever, funny, or implicit, **try reading the text out loud a few times** to take advantage of the fact that your ear is trained to notice wordplay!
- **Keep a list** of words that have multiple meanings.
  - If you want, you can start by **looking up and printing out** a list, but it is probably more valuable to **create your own list as you continue** to read and increase your vocabulary.

WEEK 3 DAY 1

OVERVIEW OF ENGLISH CONCEPTS

READING PASSAGE

From "Myths and Legends of Ancient Greece and Rome"

By E.M. Berens

(Continued from Week 2's Passages.)

We will now return to Zeus and his brothers, who, having gained a complete victory over their enemies, began to consider how the world, which they had conquered, should be divided between them. At last it was settled by lot that Zeus should reign supreme in Heaven, whilst Aides governed the Lower World, and Poseidon had full command over the Sea, but the supremacy of Zeus was recognized in all three kingdoms, in heaven, on earth (in which of course the sea was included), and under the earth. Zeus held his court on the top of Mount Olympus, whose summit was beyond the clouds; the dominions of Aides were the gloomy unknown regions below the earth; and Poseidon reigned over the sea. It will be seen that the realm of each of these gods was enveloped in mystery. Olympus was shrouded in mists, Hades was wrapped in gloomy darkness, and the sea was, and indeed still is, a source of wonder and deep interest. Hence we see that what to other nations were merely strange phenomena, served this poetical and imaginative people as a foundation upon which to build the wonderful stories of their mythology.

The division of the world being now satisfactorily arranged, it would seem that all things ought to have gone on smoothly, but such was not the case. Trouble arose in an unlooked-for quarter. The Giants, those hideous monsters (some with legs formed of serpents) who had sprung from the earth and the blood of Uranus, declared war against the triumphant deities of Olympus, and a struggle ensued, which, in consequence of Gaea having made these children of hers invincible as long as they kept their feet on the ground, was wearisome and protracted. Their mother's precaution, however, was rendered unavailing by pieces of rock being hurled upon them, which threw them down, and their feet being no longer placed firmly on their mother-earth, they were overcome, and this tedious war (which was called the Gigantomachia) at last came to an end. Among the most daring of these earth-born giants were Enceladus, Rhoetus, and the valiant Mimas, who, with youthful fire and energy, hurled against heaven great masses of rock and burning oak-trees, and defied the lightnings of Zeus.

One of the most powerful monsters who opposed Zeus in this war was called Typhon or Typhoeus. He was the youngest son of Tartarus and Gaea, and had a hundred heads, with eyes which struck terror to the beholders, and awe-inspiring voices frightful to hear. This dreadful monster resolved to conquer both gods and men, but his plans were at length defeated by Zeus, who, after a violent encounter, succeeded in destroying him with a thunderbolt, but not before he had so terrified the gods that they had fled for refuge to Egypt, where they metamorphosed themselves into different animals and thus escaped.

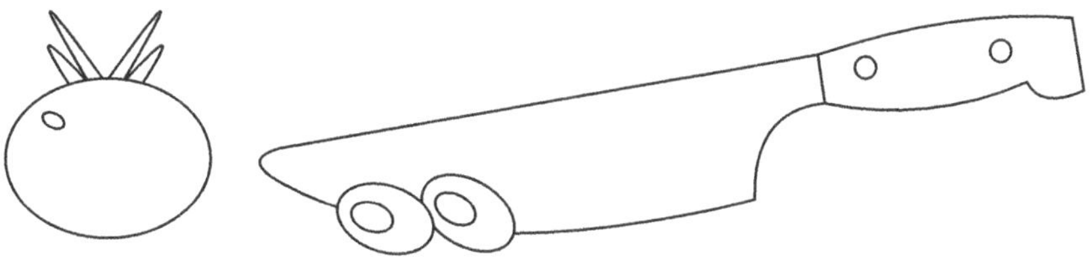

WEEK 3 DAY 1

OVERVIEW OF ENGLISH CONCEPTS
READING PASSAGE

1. How is power **shared** or **divided** between Zeus, Aides, and Poseidon?

2. Based on what we've read over the last few weeks, why is the rebellion of the **Giants** somewhat **surprising**, **ironic**, or **contrary to expectations**?

3. According to the text, why was fighting **Giants** especially <u>tricky</u>?

   A. They were huge and strong.
   B. They were invincible as long as they stood on the Earth.
   C. They were familiar with Zeus' tactics after fighting alongside him.
   D. They threw rocks down at the deities of Olympus from above.

4. How did the gods, other than Zeus, react to Typhon's attempt to conquer the world?

   A. They knew he would never succeed because Zeus had already conquered so many enemies.
   B. They joined up with Typhon because they thought he would defeat Zeus easily.
   C. They called upon the Cyclopes to help them fight again.
   D. They transformed themselves into animals to try and hide from Typhon.

5. Based on what you've read in these passages from Greek mythology, how did the Greeks feel about **power** and **violence**?

Explaining Puns & Wordplay (Pt. 1)

Directions: Review each "pun" or example of wordplay below and then, on the lines below, **explain** the wordplay to show your understanding of both the sentence's **explicit** and **implicit** meanings:

1. Our woodshop teacher talks so much that he doesn't need a drill to bore a hole in the wall.

What key word in this sentence has ***multiple meanings****, and what are those meanings?*

a. **WORD:** ____________________

b. **MEANING 1:** ____________________

____

____

c. **MEANING 2:** ____________________

____

____

What is the author's ***intended meaning*** *in this sentence? What is the* ***joke****?*

____

____

____

2. Last weekend, we went to my cousin's ranch out in the country where they keep horses. We were going to see a new pony that had just been born. When we arrived, my cousin led us over to the pen where the young colt was walking around with its mother. The older horse was making a ton of noise, but the new pony was completely silent. "He's just a little hourse," my cousin joked.

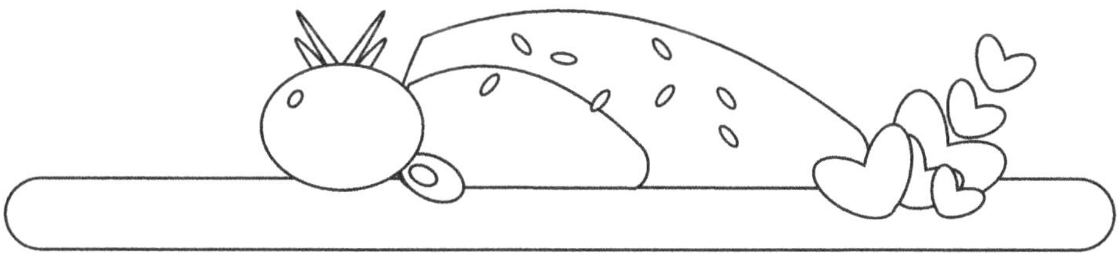

OVERVIEW OF ENGLISH CONCEPTS
ACTIVITIES

*What key word in this sentence has **a homophone**, and what are their different meanings? (**HINT:** Try reading the story out loud!)*

a. **WORD:** ____________________

b. **HOMOPHONE:** ____________________

c. **WORD MEANING:** ____________________

____

____

d. **HOMOPHONE MEANING:** ____________________

____

____

*What is the author's **intended meaning** in this sentence? What is the **joke**?*

____

____

____

FITNESS

Please be aware of your environment and be safe at all times. If you cannot do an exercise, just try your best.

Repeat these exercises **4 ROUNDS**

2 - Lunges: 8 times to each leg.
Note: Use your body weight or books as weight to do leg lunges.

1 - Abs: 20 times

3 - Plank: 20 sec.

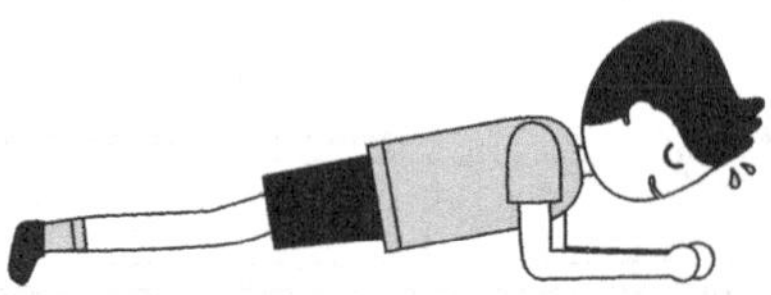

4 - Run: 50m
Note: Run **25** meters to one side and **25** meters back to the starting position.

READING PASSAGE

From Myths and Legends of Ancient Greece and Rome

By E.M. Berens

(Continued from Day 1's Passage)

Just as there were several theories concerning the origin of the world, so there were various accounts of the creation of man.

The first natural belief of the Greek people was that man had sprung from the earth. They saw the tender plants and flowers force their way through the ground in the early spring of the year after the frost of winter had disappeared, and so they naturally concluded that man must also have issued from the earth in a similar manner. Like the wild plants and flowers, he was supposed to have had no cultivation, and resembled in his habits the untamed beasts of the field, having no habitation except that which nature had provided in the holes of the rocks, and in the dense forests whose overarching boughs protected him from the inclemency of the weather.

In the course of time these primitive human beings became tamed and civilized by the gods and heroes, who taught them to work in metals, to build houses, and other useful arts of civilization. But the human race became in the course of time so degenerate that the gods resolved to destroy all mankind by means of a flood; Deucalion (son of Prometheus) and his wife Pyrrha, being, on account of their piety, the only mortals saved.

By the command of his father, Deucalion built a ship, in which he and his wife took refuge during the deluge, which lasted for nine days. When the waters abated the ship rested on Mount Othrys in Thessaly, or according to some on Mount Parnassus. Deucalion and his wife now consulted the oracle of Themis as to how the human race might be restored. The answer was, that they were to cover their heads, and throw the bones of their mother behind them. For some time they were perplexed as to the meaning of the oracular command, but at length both agreed that by the bones of their mother were meant the stones of the earth. They accordingly took up stones from the mountain side and cast them over their shoulders. From those thrown by Deucalion there sprang up men, and from those thrown by Pyrrha, women.

1. Why was it **natural** that the Greeks initially believed that people had grown from the earth?

2. Based on Paragraphs 1-2, use your own words to describe what the Greeks believed **the earliest people** were like.

3. According to the text, what caused the major flood described in Paragraph 2?

A. The gods wanted to punish the evil of humanity.
B. There was a long period of extreme weather.
C. Zeus threw a huge thunderbolt, which started a massive chain reaction.
D. Early man did not know how to cultivate the earth properly, so it led to flooding.

4. Based on the text, what is an "oracular command" (Paragraph 3)?

A. Orders to go to a certain place
B. A magic spell
C. A long journey
D. A mystical instruction obtained through religious ritual

5. According to the text, how did Pyrrha and Deucalion repopulate the world?

Explaining Puns & Wordplay (Pt. 2)

Directions: Review each "pun" or example of wordplay below. Then, on the lines below, **explain** the wordplay to show your understanding of both the sentence's **explicit** and **implicit** meanings:

1. England is known for its wet weather, which is probably due to the fact that so many kings and queens have reigned there over the years.

   *What key word in this sentence has **a homophone**, and what are their different meanings? (**HINT:** Try reading the story out loud!)*

   a. **WORD:** ______

   b. **HOMOPHONE:** ______

   c. **WORD MEANING:** ______

   d. **HOMOPHONE MEANING:** ______

   *How does the author use **homophones** for comedic effect in this sentence? What is the **joke**?*

______

WEEK 3 DAY 2

OVERVIEW OF ENGLISH CONCEPTS ACTIVITIES

2. The chef at our wedding accidentally baked the cake with salt instead of sugar, which was a gross miscalculation.

*What key word in this sentence has **multiple meanings**, and what are those meanings?*

a. **WORD:** ______________________________

b. **MEANING 1:** ______________________________

____

____

c. **MEANING 2:** ______________________________

____

____

*What is the author's **intended meaning** in this sentence? What is the **joke**?*

____

____

____

FITNESS

Please be aware of your environment and be safe at all times. If you cannot do an exercise, just try your best.

Repeat these exercises **4 ROUNDS**

1 - Squats: 20 times. Note: imagine you are trying to sit on a chair.

2 - Side Bending: 15 times to each side. Note: try to touch your feet.

3 - Tree Pose: Stay as long as possible. Note: do the same with the other leg.

MATH

Number Systems

1. -80 + (-32)

   A. -68
   B. 26
   C. -28
   D. -112

2. -47 + 21

   A. 26
   B. -26
   C. 67
   D. -67

3. 76 + (-84)

   A. -160
   B. 160
   C. -8
   D. 8

4. -83 + (-112)

   A. 195
   B. 29
   C. -195
   D. -29

5. -79 + 21

   A. -58
   B. 58
   C. 100
   D. -100

6. 75 + (-21)

   A. -54
   B. 54
   C. 96
   D. -96

7. -62 - (-34)

   A. 96
   B. -96
   C. 28
   D. -28

8. -93 - (-61)

   A. -154
   B. 154
   C. -32
   D. 32

9. -76 - 53

   A. -129
   B. 129
   C. -22
   D. 22

10. 32 - 85

   A. 53
   B. -53
   C. 113
   D. -113

WEEK 3 DAY 3 MATH

Number Systems

11. -81 - (-2)

A. 83
B. -83
C. -79
D. 79

12. -28 - 94

A. -122
B. 122
C. -66
D. 66

13. 54 - (-14)

A. -40
B. 40
C. -68
D. 68

14. -47 - 73

A. 120
B. -120
C. 26
D. -26

15. -65 - (-32)

A. 96
B. -97
C. -33
D. 33

FITNESS

Please be aware of your environment and be safe at all times. If you cannot do an exercise, just try your best.

Repeat these exercises
4 ROUNDS

1 - Bend forward: 15 times.
Note: try to touch your feet. Make sure to keep your back straight and if needed you can bend your knees.

2 - Lunges: 12 times to each leg.
Note: Use your body weight or books as weight to do leg lunges.

3 - Plank: 20 sec.

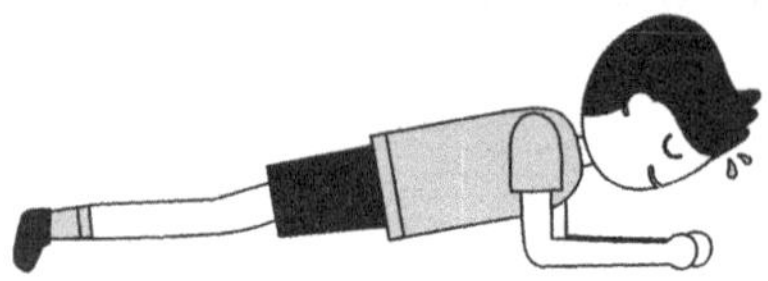

4 - Abs: 20 times

WEEK 3 DAY 4 MATH

Number Systems

1. Solve and show your work. 155+ (-739)

2. Solve and show your work. -628 + 625

3. Solve and show your work. 284 - (-847)

4. Solve and show your work. -769 - 516

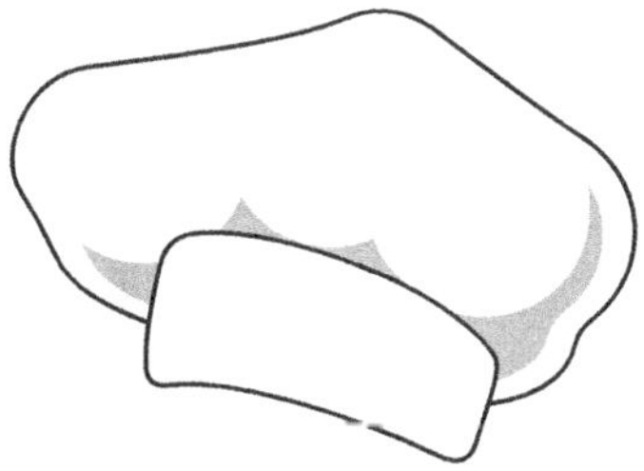

5. Solve and show your work. 937 - (-328)

6. Solve and show your work. 431 - 941

7. What is the distance between -11 and -4 on the number line?

   A. 15
   B. 7
   C. 11
   D. 4

8. What is the distance between $-4\frac{1}{2}$ and 6 on the number line?

   A. 4
   B. 2
   C. $10\frac{1}{2}$
   D. $1\frac{1}{2}$

9. Which expression represents the distance between -23 and -46 on the number line?

   A. -46 - 23
   B. 46 - (-23)
   C. -23 + (-46)
   D. -23 - (-46)

WEEK 3 DAY 4 MATH

Number Systems

10. Which expression represents the distance between 57 and -8 on the number line?

A. -57 - 8
B. -57 + 8
C. 57 - 8
D. 57 + 8

11. Which answer is undefined?

A. 8
B. 0
C. $\frac{8}{0}$
D. $\frac{0}{8}$

12. Which answer is undefined?

A. $\frac{9}{9-9}$
B. $\frac{9-9}{9}$
C. $\frac{9}{9}$
D. $\frac{1}{9}$

13. Which answer is not undefined?

A. $\frac{56}{0}$
B. $\frac{0}{24}$
C. $\frac{84}{0}$
D. $\frac{3-2}{0}$

14. Which answer is not undefined?

A. $\frac{45}{87-87}$
B. $\frac{87-87}{45}$
C. $\frac{11}{0}$
D. $\frac{32}{0}$

15. A zero in what position produces an undefined result?

A. Denominator
B. Numerator
C. Tenths place
D. Hundreds place

FITNESS

Please be aware of your environment and be safe at all times. If you cannot do an exercise, just try your best.

Repeat these exercises **4 ROUNDS**

1 - High Plank: 25 sec.

2 - Chair: 20 sec.
Note: sit on an imaginary chair, keep your back straight.

3 - Waist Hooping: 20 times.
Note: if you do not have a hoop, pretend you have an imaginary hoop and rotate your hips 10 times.

4 - Abs: 25 times

MATH

Number Systems

1. -10 ÷ (-2)

   A. 5
   B. -5
   C. 3
   D. -4

2. -21 ÷ 3

   A. -7
   B. 7
   C. -3
   D. 3

3. -20 ÷ (-5)

   A. -4
   B. -3
   C. 4
   D. -5

4. 24 ÷ (-6)

   A. -5
   B. 3
   C. 4
   D. -4

5. -88 ÷ 8

   A. 10
   B. -10
   C. 11
   D. -11

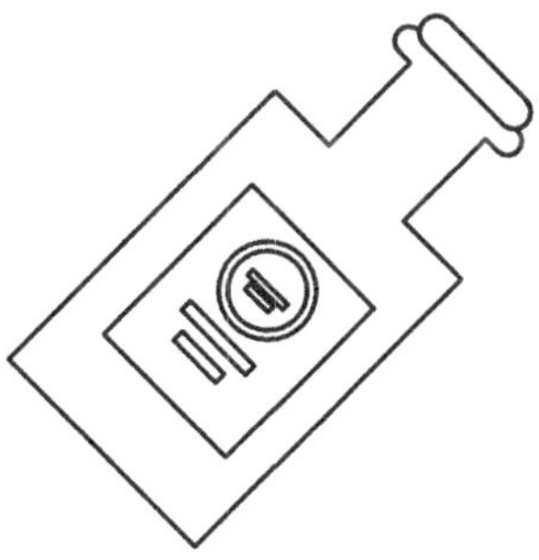

6. -6 × (-9)

   A. -54
   B. 54
   C. -63
   D. 63

7. 7 × (-3)

   A. 21
   B. -21
   C. -28
   D. 28

8. -4 × 8

   A. -32
   B. 32
   C. -24
   D. 24

9. -3 × (-6)

   A. -12
   B. 12
   C. 18
   D. -18

10. 9 × (-8)

   A. 80
   B. -80
   C. 72
   D. -72

WEEK 3 DAY 5 MATH

Number Systems

11. In the following problem, what step do you do first? 3 (8-2)+13

A. Subtract 8 and 3
B. Add 2 and 13
C. Multiply 3 and 8
D. Subtract two from eight

12. In the following problem, what step do you do first? 25÷(2x10)-11

A. Multiply 2 and 10
B. Divide 25 and 2
C. Divide 25 and 10
D. Subtract 10 from 11

13. In the following problem, what step do you do first? 56 x 3 + 11 - 14

A. 3 - 14
B. 11 - 14
C. 56 x 3
D. 3 + 11

14. In the following problem, what step do you do first? 2(4-5 +6)÷9

A. Add
B. Subtract
C. Multiply
D. Divide

15. In the following problem, what step do you do first? $3(42 \div 4^2 + 8)$

A. Square 4
B. Divide 42 and 4
C. Add 4 and 8
D. Multiply 3 and 42

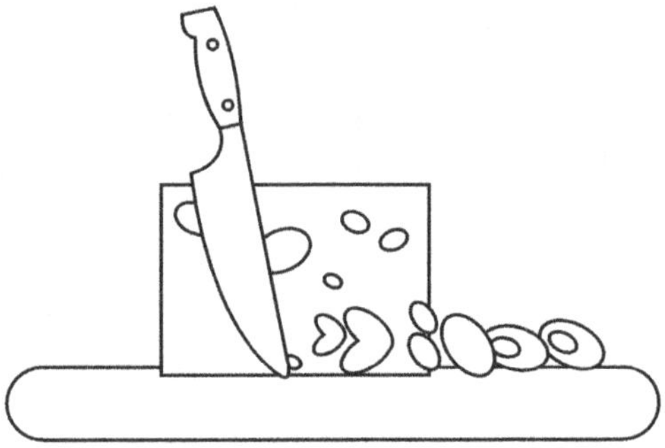

YOGA

Please be aware of your environment and be safe at all times. If you cannot do an exercise, just try your best.

1 - Down Dog: 30 sec.

2 - Bend Down: 30 sec.

3 - Chair: 30 sec.

4 - Child Pose: 30 sec.

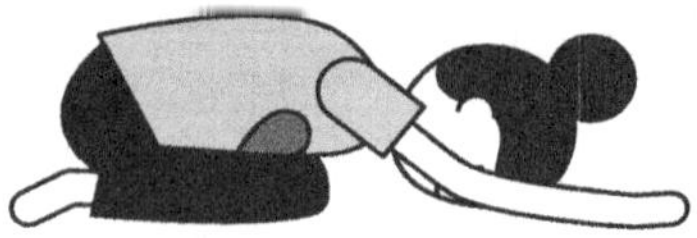

5 - Shavasana: as long as you can. Note: think of happy moments and relax your mind.

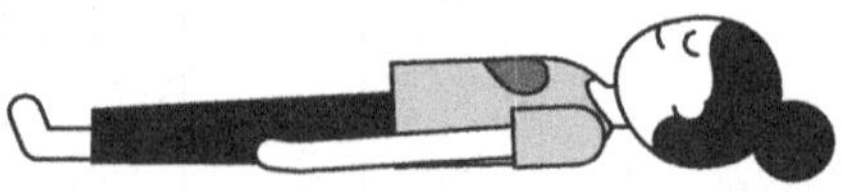

WEEK 3 DAY 6 EXPERIMENT

Newton's Third Law

Over the last two weeks, you've explored Newton's First and Second laws of motion, which state, respectively: (1) An object at rest or in motion will stay in that state unless it is acted upon by an outside force; and (2) Force = mass x acceleration. This week, we'll introduce the final of Newton's 3 Laws of Motion.

Newton's Third Law of Motion states that ***for each force, there is an equal and opposite force****. This can be harder to observe than Newton's First or Second laws because a lot of the balancing forces we're talking about are invisible to us. For example, gravity is balanced out by what's called* ***the normal force****, which represents the support that kept your book from getting pulled down through the table in Week 1's experiment.*

Today, we'll do a fun project to make Newton's Third Law visual and easy to understand!

Materials:

- 1 large inflatable balloon
- 1 drinking straw (flexible is best)
- 1 handheld toy car **or** art/crafting materials to create a small toy car rig, including...
    - Four wheels (these can be from plastic building brick kits, the hobby store, etc.)
    - Straws or pencils (to create axles)
    - Cardboard
    - Heavy tape
    - A crafting knife
- Masking tape

WEEK 3 DAY 6

EXPERIMENT

Procedure:

1. If you are creating your own car model from scratch, begin by doing that. Cut the sides, top, and bottom of the car from cardboard and assemble them using heavy tape. Attach the wheels to the car using tape, pencil axles, building bricks, or any other method that works for you. – the goal is just to create a car that can roll around on four wheels.

2. Insert the drinking straw into the open end of the uninflated balloon. Gather that rubber tightly around the straw and use masking tape to tightly secure the seal so that you can inflate the balloon by blowing into the straw.

3. Tape your balloon/straw setup to the top of your car, whether it is homemade or an existing toy. Make sure you get a good seal so the balloon stays attached.

4. Set your car on the ground or a flat tabletop. Take a second to predict what will happen when you inflate the balloon and then step away from the car. How do you think it will move? In which direction?

5. Blow into the straw to inflate the balloon. Once you are satisfied that you have a good amount of air in there, step away, carefully pinching the end of the straw so that the balloon does not deflate. When you are ready, unpinch the straw and see what happens with the car.

6. Once you've observed the car's movement, compare it to your prediction and see if you were correct or, if not, in which ways. The air that escaped the balloon actually pushed the car in the **opposite** direction, proving that **for every force, there is an equal and opposite force**. The air might have been shooting out one direction, creating a force you could feel with your hand, but an equal force pushed the car the opposite direction.

WEEK 3 DAY 6 EXPERIMENT

Follow-Up Questions:

1. What do you think would happen if you tried this experiment with a **small car that was very light**? What makes you say that?

____

____

____

2. What do you think would happen if you tried this experiment with a **car made from metal that was very heavy**? What makes you say that?

____

____

____

YOGA

Please be aware of your environment and be safe at all times. If you cannot do an exercise, just try your best.

2 - Down Dog: 30 sec.

3 - Stretching: Stay as long as possible.
Note: do on one leg then on another.

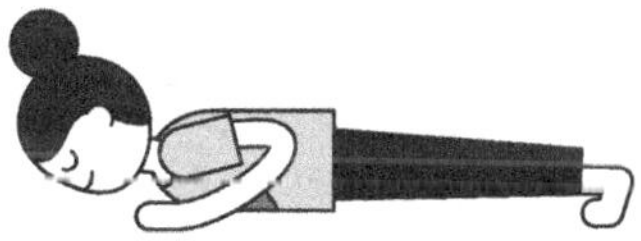

4 - Lower Plank: 20 sec.
Note: Keep your back straight and body tight.

1 - Tree Pose: Stay as long as possible. Note: do on one leg then on another.

6 - Shavasana: 5 min.
Note: this pose is very important and provides you with long term benefits. Try not to skip this. Close your eyes and imagine who you want to be and what your goals are! Always think happy thoughts.

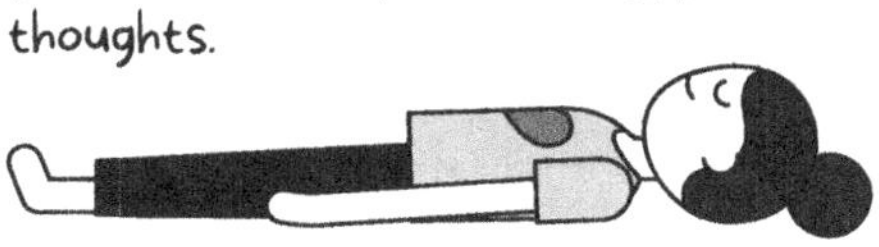

5 -Book Pose: 20 sec.
Note: Keep your core tight. Legs should be across from your eyes.

WEEK 3 DAY 7 MAZE

TASK:

From the start line (located on bottom left) go to the end (located in the center).

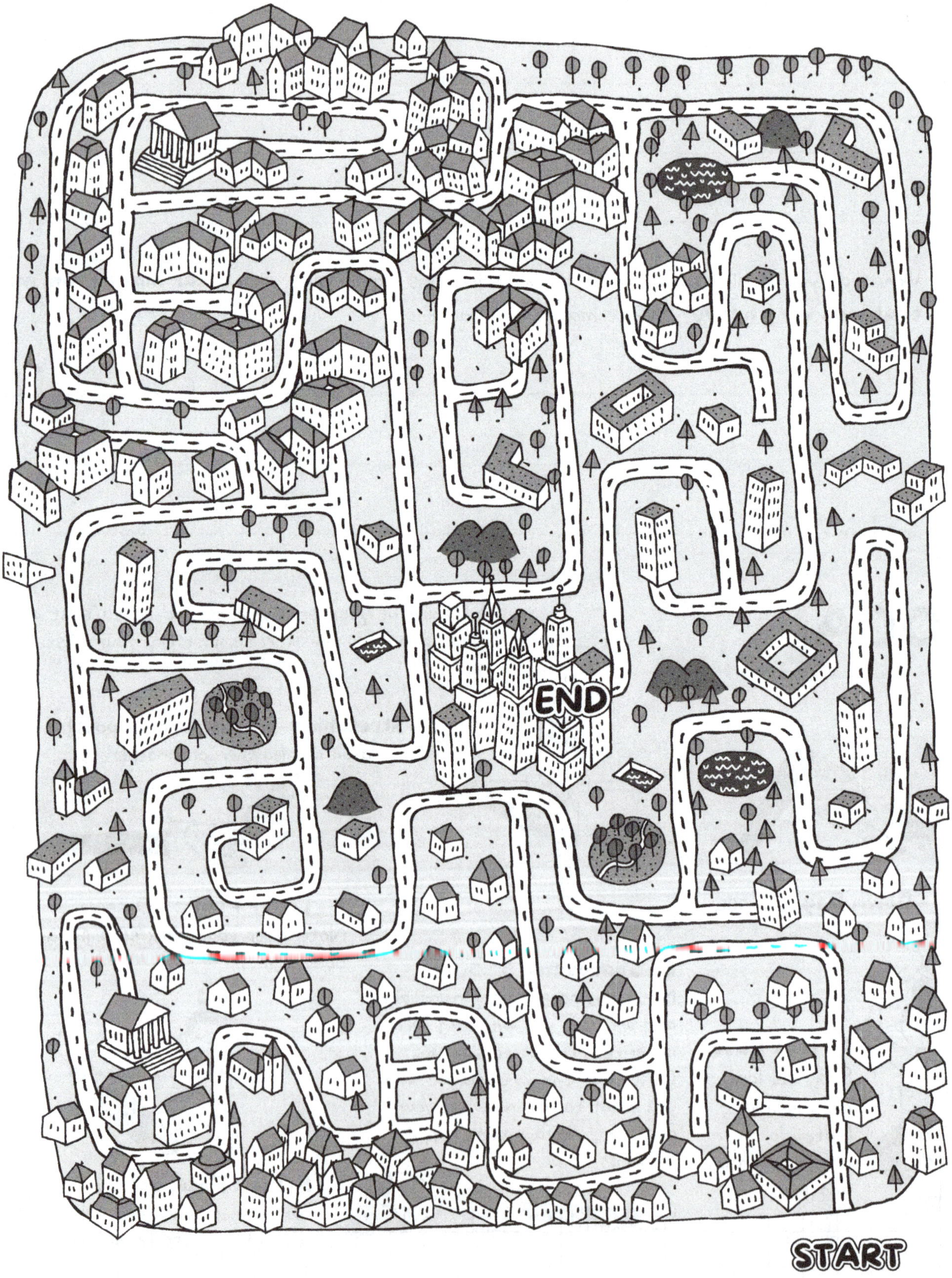

WEEK 4

Keep at it!
Do you know what an allusion is? Do you like fractions? Fear not, because this week we will be working on that and more!
25%

WEEK 4 DAY 1 — OVERVIEW OF ENGLISH CONCEPTS

Allusions

Recently, we've been discussing ways that authors build **implicit meaning** into their writing. Last week, we saw how **word play** can be used to hint at or joke about ideas within a text without ever approaching them directly. This week, we'll look at a similar tool authors can use: **allusions**.

An **allusion** is a reference within a text to something else - it could be another text, a famous person, a well-known story, etc. When it comes to an allusion, there are two main things you need to understand: **what the author is alluding to** and **what the allusion means or adds to the story.** Without a full understanding of both aspects, you won't be able to fully comprehend the **implicit subtext** of what you're reading, viewing, listening to, or experiencing.

Key Terms:

Allusion: An implicit reference to another text or idea

Why Do Writers Use Allusions?

Allusions allow an author to compare a situation to something else without needing to take the time to explain that whole story to the audience. On the other hand, an author can also use an allusion to hide certain ideas or aspects of the text from people who don't "get it." In that way, making allusions is almost like **writing in a secret code** for people who have the same reference points as you.

Throughout history, many writers have used **allusions** to connect their work to the tradition of the writers who came before them and influenced them and to show off all the different books they have read!

How Can Readers Approach Allusions?

Thanks to the internet, it's actually pretty easy to decode most **allusions** with a simple online search. With that said, really understanding why an author **chose** to make an allusion and what they are trying to achieve with it still requires some critical thinking! Let's walk through **an example** to model these thinking skills:

For example, let's say you read something with a character who, just before taking a big risk, says they are worried they are "acting like Icarus." Let's break down the steps it would take to form a strong understanding of this **allusion**:

1. Use an internet browser or encyclopedia to search for **Icarus**
    a. This would reveal that Icarus is a character from **Greek Mythology** who could fly using a pair of wax wings but died when he flew too close to the sun which melted his wings.
        i. Now you know **what the author is alluding to!**

OVERVIEW OF ENGLISH CONCEPTS

2. Ask yourself: **What is the main message or most memorable aspect of this other text or thing the author is alluding to?**
    a. In the case of Icarus, it's memorable that the character died by doing something that was risky and unwise.
3. Ask yourself: **How does the main message of the allusion connect to the text?**
    a. In our sample text, we said the character was about to take a risk. Therefore, we know that the **allusion** is being used to illustrate that the character is concerned that the risk they are about to take might be unwise.

If you use this three-step process, you'll be a master of allusions in no time!

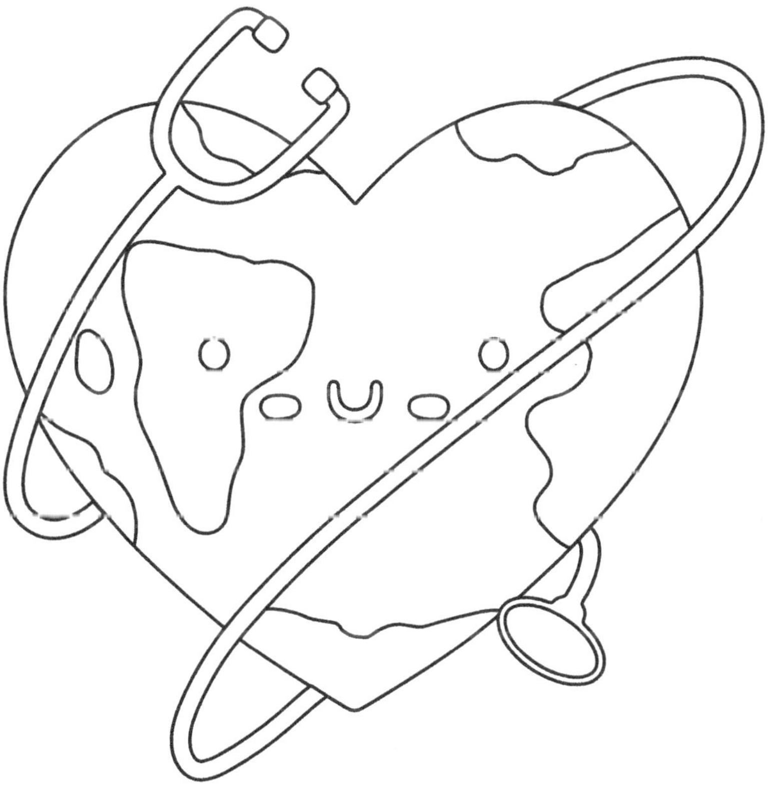

From "Pygmalion"

By Bernard Shaw

*(**NOTE:** The text we will be exploring over the next three weeks is a **play**)*

Covent Garden at 11.15 p.m. Torrents of heavy summer rain. Cab whistles blowing frantically in all directions. Pedestrians running for shelter into the market and under the portico of St. Paul's Church, where there are already several people, among them a lady and her daughter in evening dress. They are all peering out gloomily at the rain, except one man with his back turned to the rest, who seems wholly preoccupied with a notebook in which he is writing busily.

The church clock strikes the first quarter.

THE DAUGHTER [in the space between the central pillars, close to the one on her left] I'm getting chilled to the bone. What can Freddy be doing all this time? He's been gone twenty minutes.

THE MOTHER [on her daughter's right] Not so long. But he ought to have got us a cab by this.

A BYSTANDER [on the lady's right] He won't get no cab not until half-past eleven, missus, when they come back after dropping their theatre fares.

THE MOTHER. But we must have a cab. We can't stand here until half-past eleven. It's too bad.

THE BYSTANDER. Well, it ain't my fault, missus.

THE DAUGHTER. If Freddy had a bit of gumption, he would have got one at the theatre door.

THE MOTHER. What could he have done, poor boy?

THE DAUGHTER. Other people got cabs. Why couldn't he?

Freddy rushes in out of the rain from the Southampton Street side, and comes between them closing a dripping umbrella. He is a young man of twenty, in evening dress, very wet around the ankles.

THE DAUGHTER. Well, haven't you got a cab?

FREDDY. There's not one to be had for love or money.

THE MOTHER. Oh, Freddy, there must be one. You can't have tried.

READING PASSAGE

THE DAUGHTER. It's too tiresome. Do you expect us to go and get one ourselves?

FREDDY. I tell you they're all engaged. The rain was so sudden: nobody was prepared; and everybody had to take a cab. I've been to Charing Cross one way and nearly to Ludgate Circus the other; and they were all engaged.

THE MOTHER. Did you try Trafalgar Square?

FREDDY. There wasn't one at Trafalgar Square.

THE DAUGHTER. Did you try?

FREDDY. I tried as far as Charing Cross Station. Did you expect me to walk to Hammersmith?

THE DAUGHTER. You haven't tried at all.

THE MOTHER. You really are very helpless, Freddy. Go again; and don't come back until you have found a cab.

1. Which details in the text provide hints that this play is set in **England**?

____

____

2. How does the **bystander** seem different from the other characters in this scene?

____

____

____

____

3. Why is Freddy having so much trouble finding a cab?

A. A big show just ended at the theatre.
B. A big sports game just ended.
C. The characters are at a royal event.
D. The cab drivers are on strike.

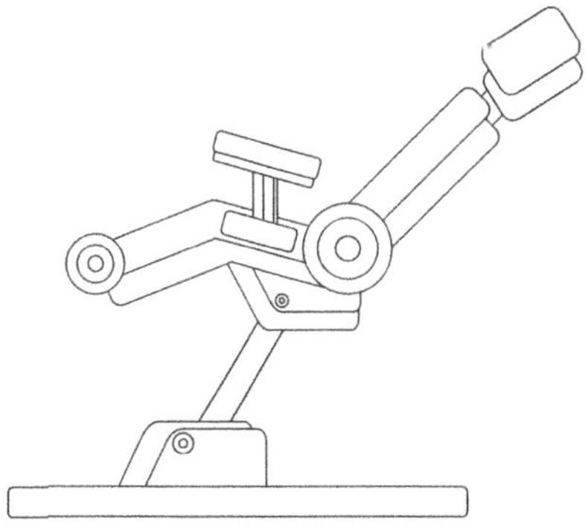

4. Which of these is the closest synonym for "engaged" as Freddy uses it when he says, "I tell you they're all engaged"?

   A. Married
   B. Occupied
   C. Closed
   D. Fighting

5. Based on this passage, what else can you **predict** about the characters of The Mother and The Daughter? How do they seem to you? What do their values and priorities seem to be?

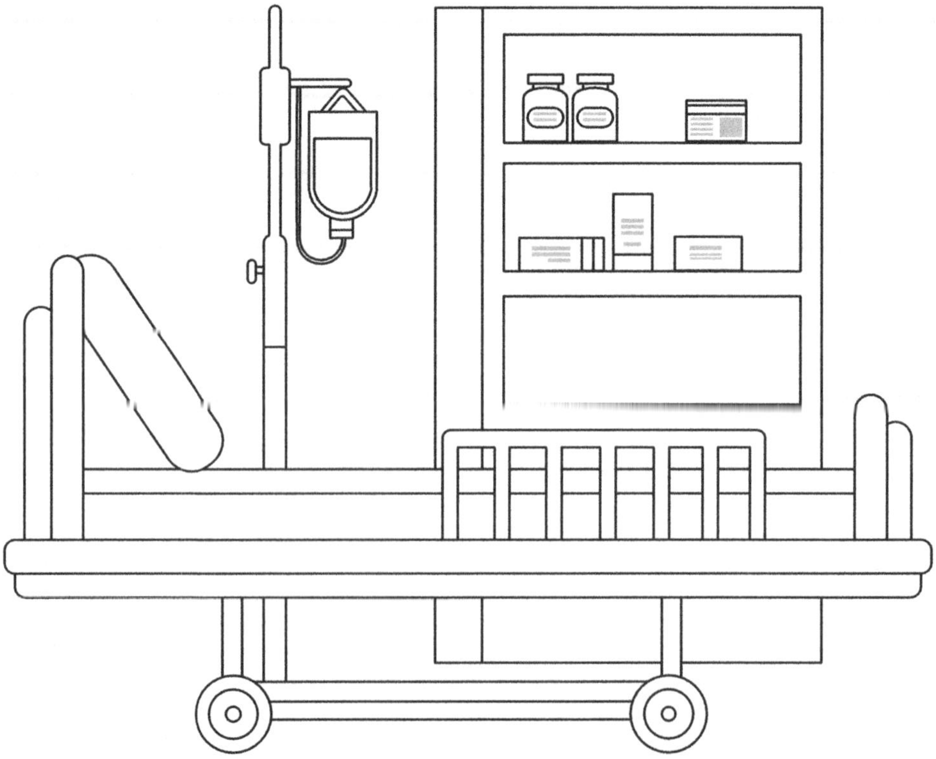

Decoding Allusions (Pt. 1)

Directions: Read each situation below with an emphasis on the underlined allusion. Using the internet for research or a set of encyclopedias, figure out **what the author is alluding to, what that allusion is all about, and how that allusion connects to or provides new insight into the situation being described.**

1. The mother applied the judgment of Solomon, telling the children that if they couldn't stop fighting over who could play the video game, she would simply return it to the store.

   a. **TO WHAT** is this author alluding? ______________________

   ____

   b. **WHAT'S THE STORY OR MAIN IDEA BEHIND THAT ALLUSION?:** ______

   ____

   ____

   ____

   ____

   c. **HOW DOES THAT ALLUSION CONNECT TO OR PROVIDE INSIGHT INTO OUR SENTENCE?:** ______________________

   ____

   ____

   ____

2. Mark and I never got along in elementary school, and now he keeps inviting me over to his house to play video games. I can't tell if he actually wants to be my friend or if this is some kind of a Trojan Horse.

   a. **TO WHAT** is this author alluding? ______________________

   ____

b. WHAT'S THE STORY OR MAIN IDEA BEHIND THAT ALLUSION?: ______

______

______

______

______

c. HOW DOES THAT ALLUSION CONNECT TO OR PROVIDE INSIGHT INTO OUR SCENARIO?: ______

______

______

______

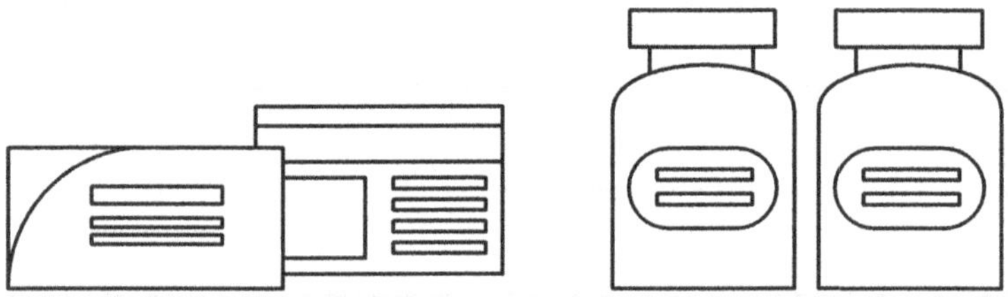

FITNESS

Please be aware of your environment and be safe at all times. If you cannot do an exercise, just try your best.

Repeat these exercises **4 ROUNDS**

2 - Lunges: 8 times to each leg.
Note: Use your body weight or books as weight to do leg lunges.

1 - Abs: 20 times

3 - Plank: 20 sec.

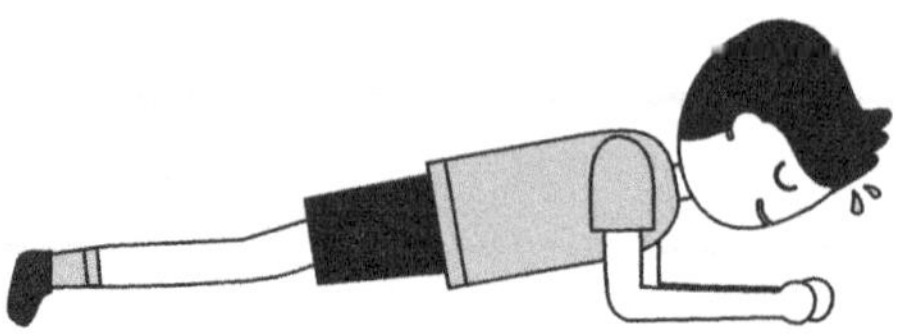

4 - Run: 50m
Note: Run 25 meters to one side and 25 meters back to the starting position.

From "Pygmalion"

By Bernard Shaw

(Continued from Day 1's passage)

THE MOTHER. You really are very helpless, Freddy. Go again; and don't come back until you have found a cab.

FREDDY. I shall simply get soaked for nothing.

THE DAUGHTER. And what about us? Are we to stay here all night in this draught, with next to nothing on. You selfish pig -

FREDDY. Oh, very well: I'll go, I'll go. [He opens his umbrella and dashes off Strandwards, but comes into collision with a flower girl, who is hurrying in for shelter, knocking her basket out of her hands. A blinding flash of lightning, followed instantly by a rattling peal of thunder, orchestrates the incident]

THE FLOWER GIRL. Nah then, Freddy: look wh' y' gowin, deah. {**TRANSLATION:** "Now then, Freddy: look where you're going, will you?"}

FREDDY. Sorry [he rushes off].

THE FLOWER GIRL [picking up her scattered flowers and replacing them in the basket] There's menners f' yer! Te-oo banches o voylets trod into the mad. {**TRANSLATION:** "There's manners for you! Two bunches of violets trod into the mud!"}

[She sits down on the plinth of the column, sorting her flowers, on the lady's right. She is not at all an attractive person. She is perhaps eighteen, perhaps twenty, hardly older. She wears a little sailor hat of black straw that has long been exposed to the dust and soot of London and has seldom if ever been brushed. Her hair needs washing rather badly. its mousy color can hardly be natural. She wears a shoddy black coat that reaches nearly to her knees and is shaped to her waist. She has a brown skirt with a coarse apron. Her boots are much the worse for wear. She is no doubt as clean as she can afford to be; but compared to the ladies she is very dirty. Her features are no worse than theirs; but their condition leaves something to be desired; and she needs the services of a dentist].

THE MOTHER. How do you know that my son's name is Freddy, pray?

THE FLOWER GIRL. Ow, eez ye-ooa san, is e? Wal, fewd dan y' de-ooty bawmz a mather should, eed now bettern to spawl a pore gel's flahrzn than ran awy atbaht pyin. Will ye-oo py me f'them? {**TRANSLATION:** "Oh, he's your son, is he? Well, if you'd done your duty like a mother should, he'd know better than to spoil a girl's flowers then run away without paying for them. Will you pay me for them?"}

READING PASSAGE

THE DAUGHTER. Do nothing of the sort, mother. The idea!

THE MOTHER. Please allow me, Clara. Have you any pennies?

THE DAUGHTER. No. I've nothing smaller than sixpence.

THE FLOWER GIRL [hopefully] I can give you change for a tanner, kind lady.

THE MOTHER [to Clara] Give it to me. [Clara parts reluctantly]. Now [to the girl] This is for your flowers.

THE FLOWER GIRL. Thank you kindly, lady.

THE DAUGHTER. Make her give you the change. These things are only a penny a bunch.

THE MOTHER. Do hold your tongue, Clara. [To the girl]. You can keep the change.

THE FLOWER GIRL. Oh, thank you, lady.

THE MOTHER. Now tell me how you know that young gentleman's name.

THE FLOWER GIRL. I didn't.

THE MOTHER. I heard you call him by it. Don't try to deceive me.

THE FLOWER GIRL [protesting] Who's trying to deceive you? I called him Freddy or Charlie same as you might yourself if you was talking to a stranger and wished to be pleasant. [She sits down beside her basket].

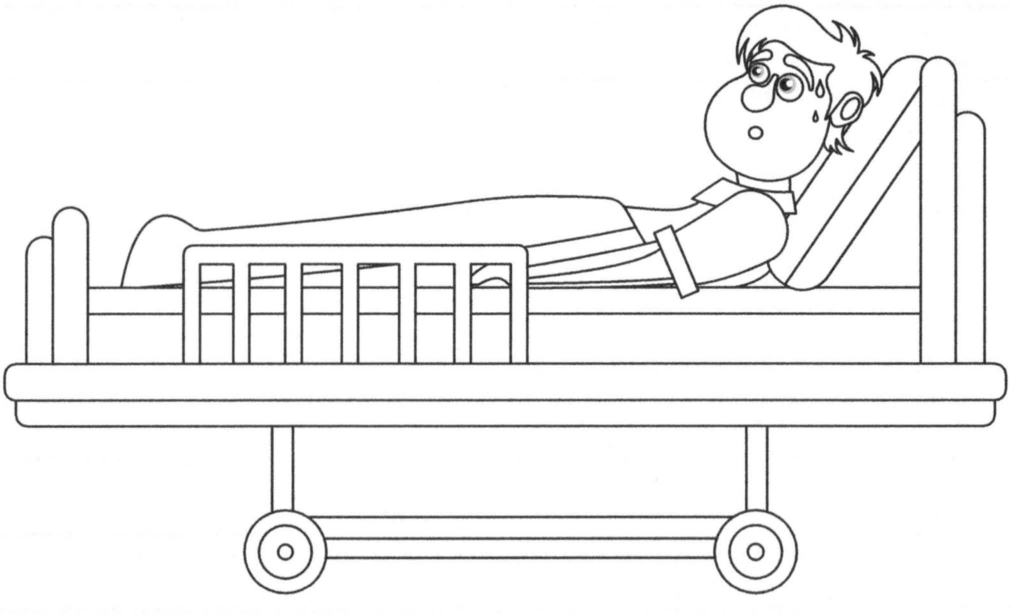

READING PASSAGE

1. How does Freddy prove his mother right for saying that he was "hapless" (unlucky) in the previous scene?

2. How does the author make it easy to tell that **The Flower Girl** is going to be an important character in this play?

3. How does the Flower Girl explain that she knows Freddy's name?

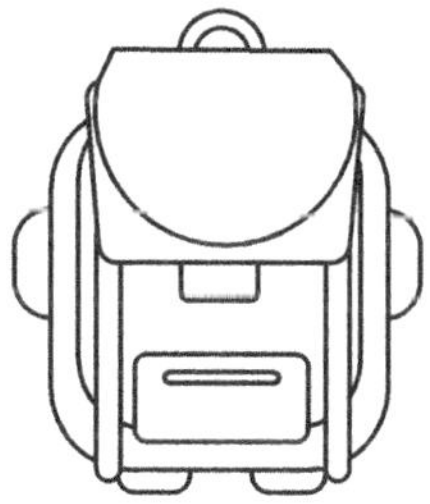

   A. He has bought flowers from her in the past.
   B. They used to know each other when they were children.
   C. She just used "Freddy" as a generic name for a young man.
   D. She had heard the family talking to him before.

4. How are The Mother and The Daughter fundamentally different in this scene?

   A. The Mother is kind to Freddy, but the Daughter is rude to him.
   B. The Daughter is kind to Freddy, but the Mother is rude to him.
   C. The Mother is decent to the Flower Girl, but the Daughter is rude to her.
   D. The Mother is rude to the Flower Girl, but the daughter is decent to her.

5. Imagine this scene is staged and performed as a drama. What aspects of this scene do you imagine are **funny** when it is performed live by actors for an audience?

Decoding Allusions (Pt. 2)

Directions: Read each situation below with an emphasis on searching for allusions (they will **not** be underlined this time)! Once you've found the allusion, use the internet for research, or a set of encyclopedias,to figure out **what the author is alluding to, what that allusion is all about**, and **how that allusion connects to or provides new insight into the situation being described.**

1. I hid my dad's watch under the couch cushion to play a prank on him, but he found it right away because the old thing was ticking like the Tell-Tale Heart.

   a. **TO WHAT** is this author alluding? ______________________________

   ____

   b. **WHAT'S THE STORY OR MAIN IDEA BEHIND THAT ALLUSION?:** ______

   ____

   ____

   ____

   ____

   c. **HOW DOES THAT ALLUSION CONNECT TO OR PROVIDE INSIGHT INTO OUR SENTENCE?:** ______________________________

   ____

   ____

   ____

2. After my back surgery, I was walking around so slowly and stiffly that I started checking in the mirror to see if the doctors had put bolts in my neck as well.

   a. **TO WHAT** is this author alluding? ______________________________

   ____

   b. **WHAT'S THE STORY OR MAIN IDEA BEHIND THAT ALLUSION?:** ______

   ____

   ____

   ____

   ____

   c. **HOW DOES THAT ALLUSION CONNECT TO OR PROVIDE INSIGHT INTO OUR SCENARIO?:** ______________________________

   ____

   ____

   ____

FITNESS

Please be aware of your environment and be safe at all times. If you cannot do an exercise, just try your best.

Repeat these exercises **4 ROUNDS**

1 - Squats: 20 times.
Note: imagine you are trying to sit on a chair.

2 - Side Bending: 15 times to each side. Note: try to touch your feet.

3 - Tree Pose: Stay as long as possible.
Note: do the same with the other leg.

WEEK 4 DAY 3 MATH

Number Systems

1. Simplify $\frac{15}{30}$

A. $\frac{1}{3}$

B. $\frac{10}{25}$

C. $\frac{1}{2}$

D. $\frac{2}{1}$

2. Simplify $\frac{14}{42}$

A. $\frac{1}{3}$

B. $\frac{3}{1}$

C. $\frac{4}{30}$

D. $\frac{1}{6}$

3. Simplify $\frac{6m}{15m}$

A. $\frac{1}{7}$

B. $\frac{6}{15}$

C. $\frac{2}{5}$

D. $\frac{2m}{5m}$

4. Simplify $\frac{18x}{32x^2}$

A. $\frac{9}{16}$

B. $\frac{9}{16x}$

C. $\frac{9x}{16}$

D. $\frac{9}{6x}$

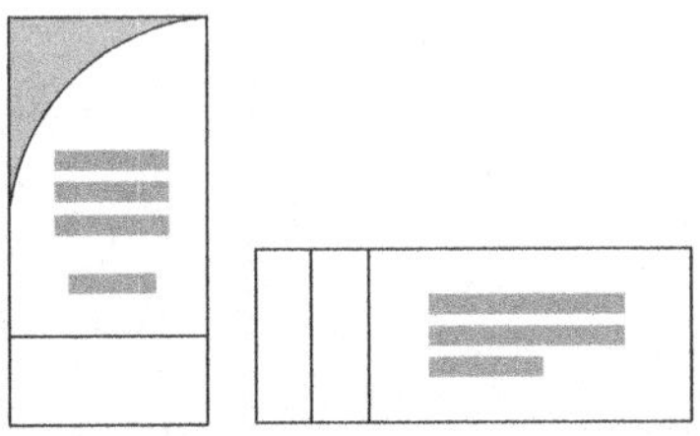

5. Simplify $\frac{40ab^2}{5ab}$

A. $8a$

B. $8b$

C. $\frac{1}{8b}$

D. $\frac{1}{8a}$

6. Which fraction is equal to $\frac{-5}{8}$?

A. $-\frac{-5}{8}$

B. $\frac{-5}{-8}$

C. $\frac{5}{8}$

D. $-\frac{5}{8}$

7. Which fraction is equal to $\frac{-1}{-4}$?

A. $-\frac{1}{4}$

B. $\frac{1}{-4}$

C. $\frac{1}{4}$

D. $\frac{-1}{4}$

8. Which fraction is equal to $\frac{7}{-9}$?

A. $\frac{-7}{9}$

B. $\frac{-7}{-9}$

C. $\frac{7}{--9}$

D. $\frac{--7}{9}$

WEEK 4 DAY 3

MATH

9. Which fraction is equal to $\frac{-12b}{16b}$?

A. $\frac{3b}{4}$

B. $-\frac{3b}{4}$

C. $-\frac{3}{4}$

D. $\frac{3}{4}$

10. Which fraction is equal to $\frac{56j}{-42j^2}$?

A. $-\frac{6j}{8}$

B. $-\frac{8}{-6j}$

C. $-\frac{3j}{4}$

D. $\frac{-8}{6j}$

11. Which number is the greatest?

A. $\frac{9}{10}$

B. -9

C. $\frac{-11}{10}$

D. $\frac{1}{10}$

12. Which number is the greatest?

A. $\frac{5}{2}$

B. $\frac{3}{2}$

C. $\frac{1}{3}$

D. $-\frac{5}{2}$

13. Which number is the least?

A. $\frac{1}{8}$

B. $\frac{3}{2}$

C. 4

D. -6

14. Which number is the least?

A. $-\frac{1}{3}$

B. -4

C. -14

D. -8

15. Which number is the least?

A. -113

B. -121

C. -118

D. -105

FITNESS

Please be aware of your environment and be safe at all times. If you cannot do an exercise, just try your best.

Repeat these exercises **4 ROUNDS**

1 - Bend forward: 15 times.
Note: try to touch your feet. Make sure to keep your back straight and if needed you can bend your knees.

2 - Lunges: 12 times to each leg.
Note: Use your body weight or books as weight to do leg lunges.

3 - Plank: 20 sec.

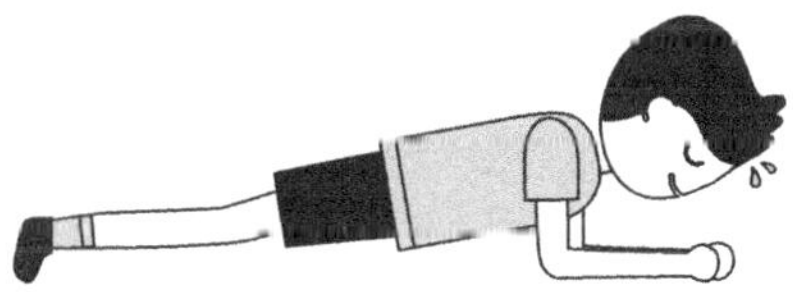

4 - Abs: 20 times

Order of operations

1. $4(4) - 2(5)$

   A. 6
   B. 26
   C. 16
   D. 40

2. $7 + 3\ (8 \div 4)$

   A. 20
   B. 13
   C. 15
   D. 24

3. $35 \div 7 - 2$

   A. 7
   B. 3
   C. 5
   D. 9

4. $21 \div 7 + 14 \times 2$

   A. 24
   B. 27
   C. 29
   D. 31

5. $2 + 3 \times 24 \div 6$

   A. 10
   B. 12
   C. 14
   D. 16

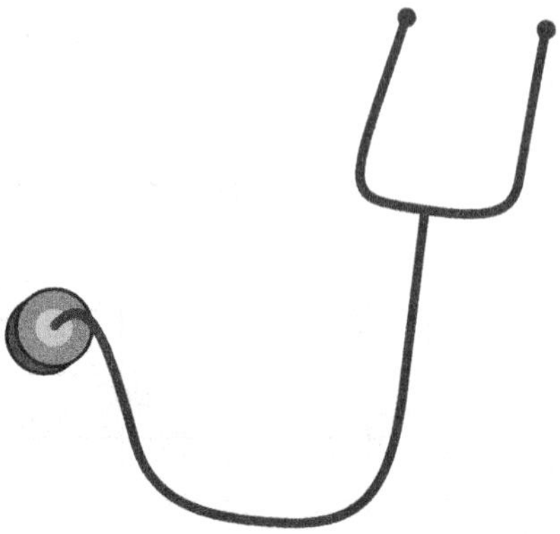

6. $2(15 - 9) \times 9$

   A. 75
   B. 6
   C. 54
   D. 108

7. $3(7 + 4)$

   A. 33
   B. 25
   C. 30
   D. 22

8. $\frac{7 + 8}{3} + 2(6)$

   A. 17
   B. 15
   C. 13
   D. 11

9. $\frac{21 + 15}{3 + 6}$

   A. 1
   B. 2
   C. 3
   D. 4

10. $8 - 2 \times 3$

   A. 7
   B. 14
   C. 2
   D. 18

WEEK 4 DAY 4 MATH

11. $6 \times 2 + 4$

A. 12
B. 14
C. 16
D. 36

12. $[22 + (0.4 \times 2)] - 3$

A. 12.9
B. 19.8
C. 41.4
D. 41.8

13. $4 \div 4 \times 4 + 4 - 4$

A. 4
B. 8
C. 0
D. 16

14. $[2 + (6 \times 8)] - 1$

A. 46
B. 48
C. 49
D. 50

15. $(21 + 3) \div 4 \div 2$

A. 24
B. 12
C. 6
D. 3

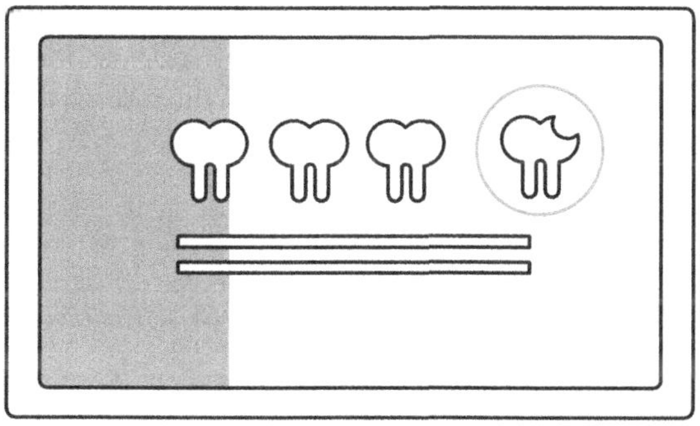

FITNESS

Please be aware of your environment and be safe at all times. If you cannot do an exercise, just try your best.

Repeat these exercises **4 ROUNDS**

1 - High Plank: 25 sec.

2 - Chair: 20 sec.
Note: sit on an imaginary chair, keep your back straight.

3 - Waist Hooping: 20 times. Note: if you do not have a hoop, pretend you have an imaginary hoop and rotate your hips 10 times.

4 - Abs: 25 times

WEEK 4 DAY 5 MATH

Order of operations

1. Use < or > or = to complete the statement:

-8 ______ 0

Answer: ____________________

2. Use < or > or = to complete the statement:

4 ______ -25

Answer: ____________________

3. Use < or > or = to complete the statement:

-9 ______ -2

Answer: ____________________

4. Use < or > or = to complete the statement:

$|-1|$ ______ $|50|$

Answer: ____________________

5. Use < or > or = to complete the statement:

$|-6|$ ______ $|-12|$

Answer: ____________________

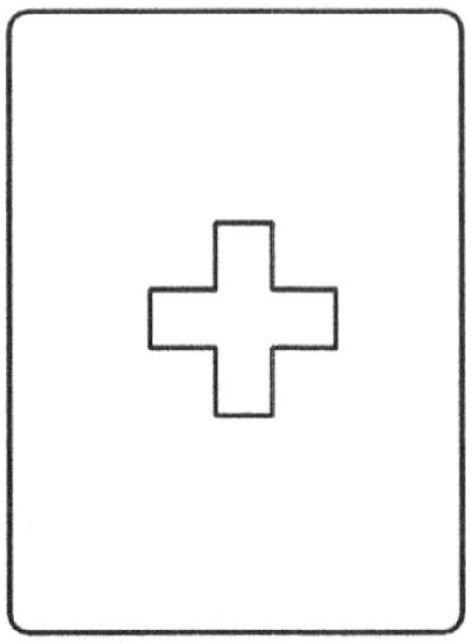

6. Use < or > or = to complete the statement:

$|10|$ ______ $|-10|$

Answer: ____________________

7. Which expression has a value of -90?

A. 90
B. $|90|$
C. $-|-90|$
D. $|-90|$

8. Which list shows the values in order from least to greatest?

A. $-6, |-3|, 3, |-6|$
B. $|-6|, 6, |-3|, 3$
C. $|-6|, - -3, |-3|, 6$
D. $-3, -|-6|, -|-3|, 6$

9. Which two integers are between -5 and 2?

A. 0, 4
B. -6, 1
C. -3, 3
D. -4, 1

10. Write the number that represents the quantity: 18° below 0

Answer: ____________________

WEEK 4
DAY 5

MATH

11. Write the number that represents the quantity: 45 sec before launch

Answer: ______________________

12. Write the number that represents the quantity: a debt of $50

Answer: ______________________

13. Write the number that represents the quantity: win by 7 points

Answer: ______________________

14. Write the number that represents the quantity: a loss of 8 yards

Answer: ______________________

15. Write the number that represents the quantity: 300 feet below sea level

Answer: ______________________

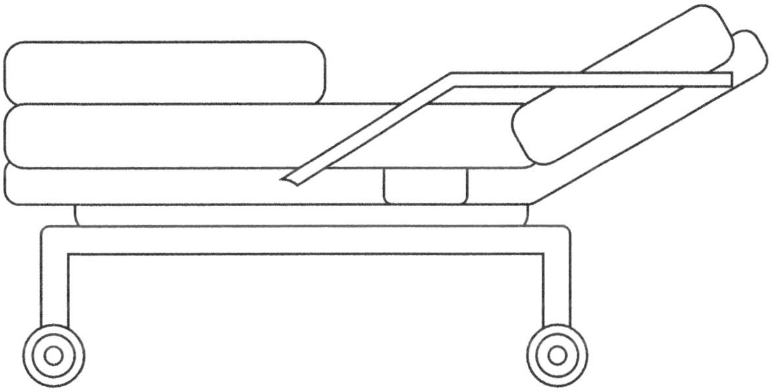

YOGA

Please be aware of your environment and be safe at all times. If you cannot do an exercise, just try your best.

1 - Down Dog: 30 sec.

2 - Bend Down: 30 sec.

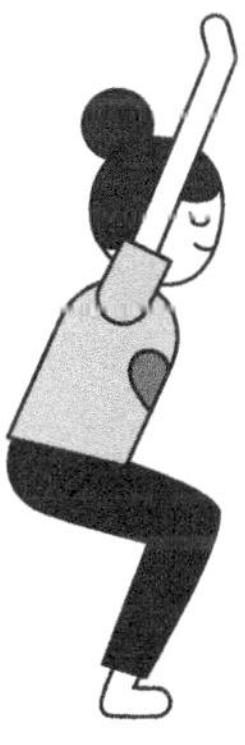

3 - Chair: 30 sec.

4 - Child Pose: 30 sec.

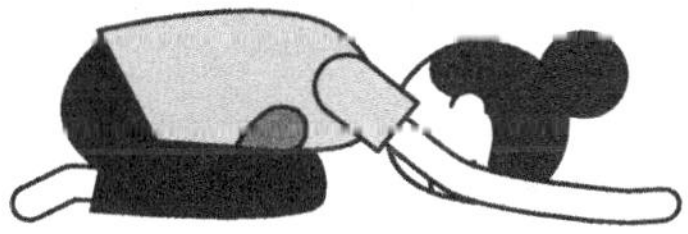

5 - Shavasana: as long as you can. Note: think of happy moments and relax your mind.

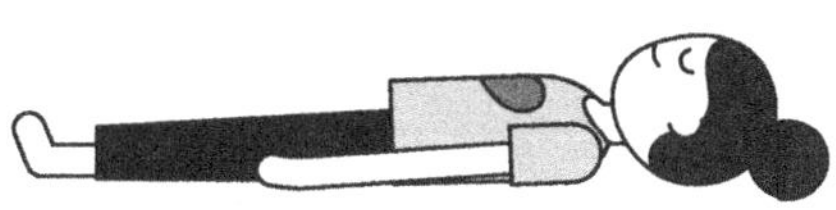

WEEK 4 DAY 6 EXPERIMENT

Visualizing Particle Motion

*Over the last few weeks, we introduced principles of **physics** and thought a lot about **motion**: Where does motion come from? What's the relationship between force and motion? What kind of forces are around us? Today, we'll be applying the same thinking on a much **smaller scale**. So far, we've looked at how **large objects** like books and cars move, but today it's time to think tiny!*

***Particles** are tiny pieces of matter that can be visible (like a spark or a grain of salt) or so small they can't be seen (like a single electron inside of an atom). Because they are so small, particles can move around at extremely high speeds. In this activity, we'll be studying the relationship between **motion** and **temperature** by studying the movement of particles.*

Materials:

- 3 clear drinking glasses
- Hot and cold running water
- A kettle (optional)
- Notepaper
- Food coloring (any color is fine)
- A few ice cubes
- Paper towel

WEEK 4 DAY 6 EXPERIMENT

Procedure:

1. Cover a tray or countertop with a few pieces of paper towel.

2. Fill one glass with the **coldest** possible water from your sink and add a few ice cubes. Give the water a few seconds to get nice and icy cold. You can even put it in the freezer for a few minutes, if you'd like, but don't let the liquid water freeze. Place the ice cold glass on the paper towel-covered tray or countertop.

3. Fill one glass with **room temperature** water from your sink. Place the glass of room temperature water on the paper towel-covered tray or countertop.

4. Fill the final glass with hot water. You can use the hottest possible water from your sink (let it run for a few seconds) or warm some water up in a kettle (although **do not use boiling water**, as that may crack your glass and create a dangerous situation). Place the glass of hot water on the paper towel-covered tray or countertop.

5. Once all three glasses are prepared, get out your notepaper and make a prediction about what will happen when a drop of food coloring is added to each glass. How do you think the food coloring will behave differently in the three glasses?

6. Add a single drop of food coloring to each glass without stirring and observe the way the particles of food coloring spread, flow, and expand throughout the glass. Take some notes describing the different ways the food coloring behaves in each glass.

7. Once you've observed all three glasses for a few minutes, answer the questions below and clean up your materials.

WEEK 4 DAY 6 EXPERIMENT

Follow-Up Questions:

1. Based on what you saw, how does **temperature** affect particle **movement**?

____

____

____

2. Brainstorm and write down an example that you can observe in daily life that shows that **cold temperatures** make **movement** more difficult:

____

____

____

YOGA

Please be aware of your environment and be safe at all times. If you cannot do an exercise, just try your best.

1 - Tree Pose: Stay as long as possible.
Note: do on one leg then on another.

2 - Down Dog: 30 sec.

3 - Stretching: Stay as long as possible.
Note: do on one leg then on another.

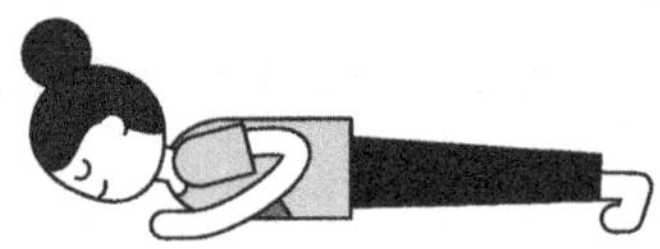

4 - Lower Plank: 20 sec.
Note: Keep your back straight and body tight.

5 -Book Pose: 20 sec.
Note: Keep your core tight. Legs should be across from your eyes.

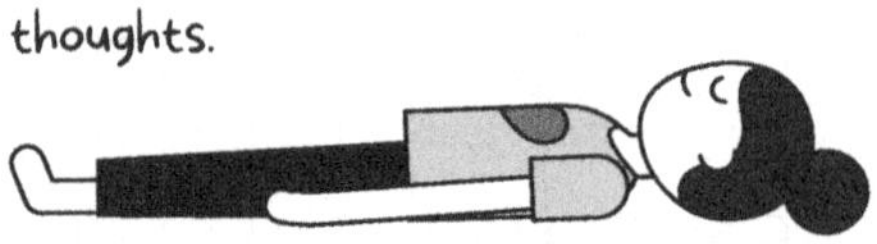

6 - Shavasana: 5 min.
Note: this pose is very important and provides you with long term benefits. Try not to skip this. Close your eyes and imagine who you want to be and what your goals are! Always think happy thoughts.

WEEK 4
DAY 7

MEMORY

Task:

Let's play a memory game! Look at the first diagram. Use a pen to shade in the same squares on the second diagram. Now cover the first two diagrams and try to remember how the diagram was shaded in. You can use the third and fourth diagrams which are empty to shade in by memory.

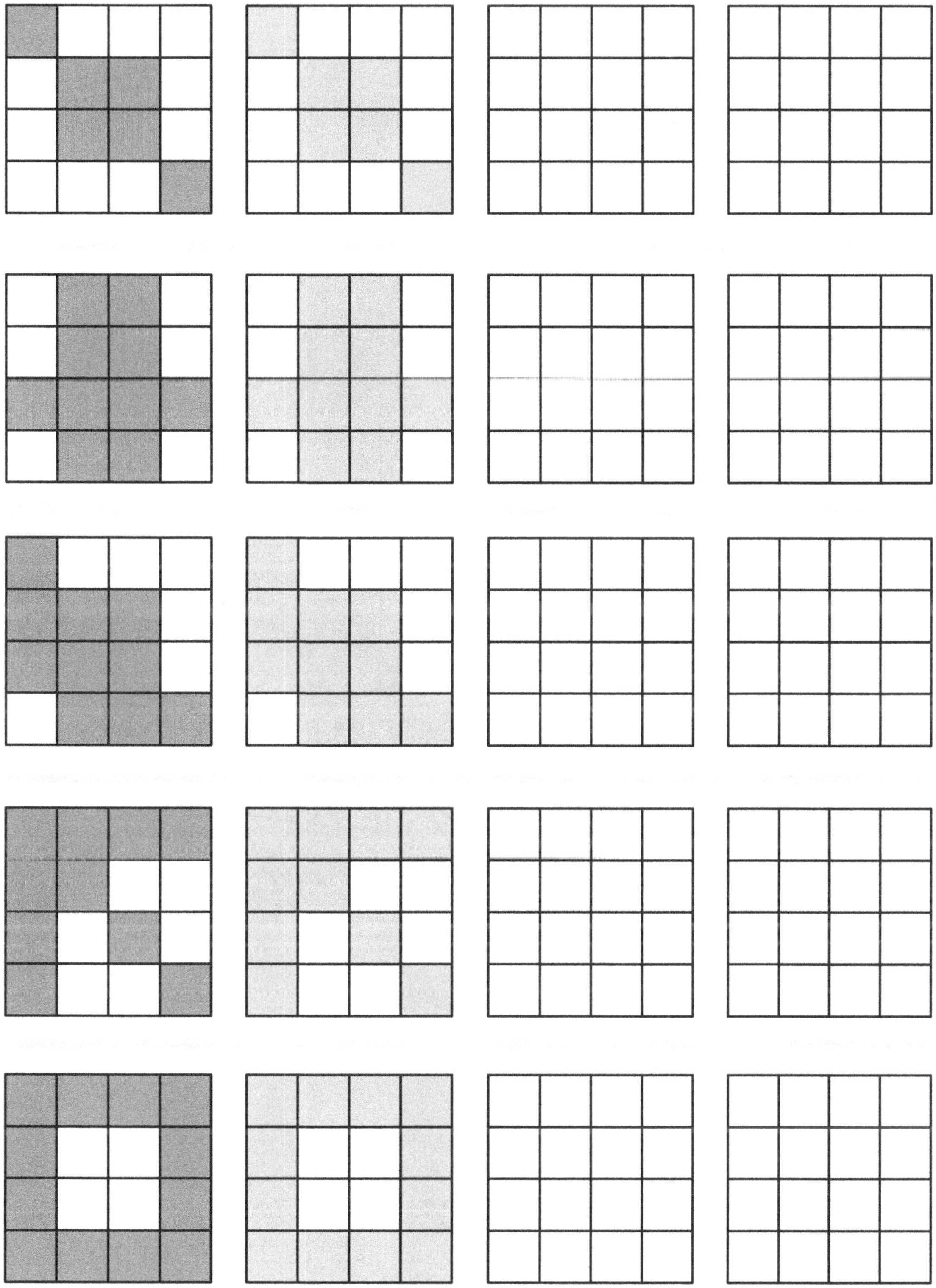

WEEK 5

Fantastic!
In 8th grade, a big topic covered is Expressions & Equations. In Week 5, we will introduce various problems that deal with that topic.

ARGOPREP

OVERVIEW OF ENGLISH CONCEPTS

Narrative Structure

We've spent a lot of time so far this summer talking about what's **not** explicitly stated in a text: the **implicit** subtext. This week, however, we're going to change things up and talk about one of the most fundamental, **explicit** parts of a literary text: the narrative structure.

In a literary text, **narrative structure** refers to both the content of the story (the **plot**) as well as the **pacing** of the text (the rate at which events unfold and characters grow or experience new things). **Generally**, most stories follow a relatively similar structure. That is not to say that all stories are alike, but rather, that the general structure and pacing of a story that "works" has been refined over time. Over the next few days, we'll explore that structure and see how it can help us better understand what we read while also improving our ability to write!

Stages of Narrative Structure

1. **Exposition:** The part at the very beginning of the story in which the **setting** and **the most essential characters** are introduced. The goal is to orient the reader.
2. **The Inciting Incident:** The part where something happens that kicks off a particular story. Usually, it involves something being wrong that needs to be fixed, something far away that needs to be gotten, etc.
3. **Rising Action:** The main character leaves the place where the story began and faces a series of increasingly tough obstacles, challenges, or opponents. At each step of the way, the **intensity** of the story goes up and the main character has to think and act in new and better ways to continue to succeed.
4. **The Climax:** The most important, decisive moment in the story when the main character breaks through and achieves something great or overcomes the greatest challenge of all. Often, the main character **changes** (physically or mentally) at the **climax**.
5. **Falling Action:** The story unwinds in intensity as things fall into the right place and the reader gets a sense of how the world is different due to or after the events of the story.
6. **Resolution:** (Usually, this part of the story is very short.) The final portion of the story that wraps everything up and hints what the main characters' lives will be like moving forward.

How Can We Use this Understanding of Narrative Structure?

- **As a reader,** understanding this structure will help you better understand "where you are" at any given point in a story. Once you understand the concept of the "inciting incident" and "rising action," you'll have a better sense of when the story is about to pick up in terms of action or when really important things are going to happen.
    - Studying the way an author structures their narrative also provides great insights into the author's personal style.
- **As a writer**, understanding plot structure is fundamental to writing narratives of any kind! Using this basic structure, an infinite number of stories can be crafted and written without any of them needing to be repetitive or boring. Structure is **truly** a writer's best friend!

READING PASSAGE

From "Pygmalion"

By Bernard Shaw

*(**NOTE:** These stage directions open the next scene in the play you began reading last week.)*

Next day at 11 a.m. Higgins's laboratory in Wimpole Street. It is a room on the first floor, looking on the street, and was meant for the drawing-room. The double doors are in the middle of the back hall; and persons entering find in the corner to their right two tall file cabinets at right angles to one another against the walls. In this corner stands a flat writing-table, on which are a phonograph, a laryngoscope, a row of tiny organ pipes with a bellows, a set of lamp chimneys for singing flames with burners attached to a gas plug in the wall by a rubber tube, several tuning-forks of different sizes, a life-size image of half a human head, showing in section the vocal organs, and a box containing a supply of wax cylinders for the phonograph.

Further down the room, on the same side, is a fireplace, with a comfortable leather-covered easy-chair at the side of the hearth nearest the door, and a coal-scuttle. There is a clock on the mantelpiece. Between the fireplace and the phonograph table is a stand for newspapers.

On the other side of the central door, to the left of the visitor, is a cabinet of shallow drawers. On it is a telephone and the telephone directory. The corner beyond, and most of the side wall, is occupied by a grand piano, with the keyboard at the end furthest from the door, and a bench for the player extending the full length of the keyboard. On the piano is a dessert dish heaped with fruit and sweets, mostly chocolates.

The middle of the room is clear. Besides the easy chair, the piano bench, and two chairs at the phonograph table, there is one stray chair. It stands near the fireplace. On the walls, engravings; mostly Piranesis and mezzotint portraits. No paintings.

Pickering is seated at the table, putting down some cards and a tuning-fork which he has been using. Higgins is standing up near him, closing two or three file drawers which are hanging out. He appears in the morning light as a robust, vital, appetizing sort of man of forty or thereabouts, dressed in a professional-looking black frock-coat with a white linen collar and black silk tie. He is of the energetic, scientific type, heartily, even violently interested in everything that can be studied as a scientific subject, and careless about himself and other people, including their feelings. He is, in fact, but for his years and size, rather like a very impetuous baby "taking notice" eagerly and loudly, and requiring almost as much watching to keep him out of unintended mischief. His manner varies from genial bullying when he is in a good humor to stormy petulance when anything goes wrong; but he is so entirely frank and void of malice that he remains likeable even in his least reasonable moments.

OVERVIEW OF ENGLISH CONCEPTS
READING PASSAGE

1. How is the setting of this scene **different** from the setting of the play's first scene, which you read last week?

2. Based on the items described, what are some of **Higgins'** interests?

3. Which of these objects is **not** in Higgins' laboratory?

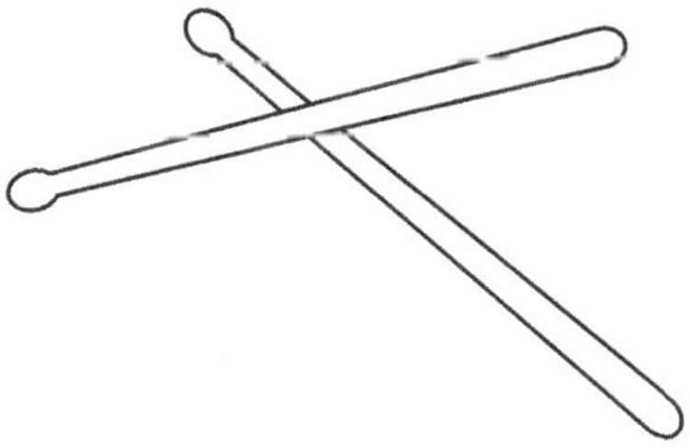

   A. Candy
   B. A Telephone
   C. A newspaper
   D. A painting

4. Which of these things does the author **not** compare Higgins to in the final paragraph?

   A. A bully
   B. A baby
   C. A professional
   D. A caring person

5. Based on this brief description of his personality and office, do you think **Higgins** is a person you'd want to meet in real life? Does he seem like someone who'd be interesting or someone who'd be difficult to deal with? What makes you say that?

OVERVIEW OF ENGLISH CONCEPTS ACTIVITIES

Breaking Down Narrative Structure

Directions: Select one of your favorite **narrative texts**. It could be a book, a short story, a movie, an episode of a TV show, etc. Think about that text with a focus on the **plot** and **narrative structure** and fill out the plotline below to show how the story builds in excitement and tension.

TEXT TITLE: ______________________________

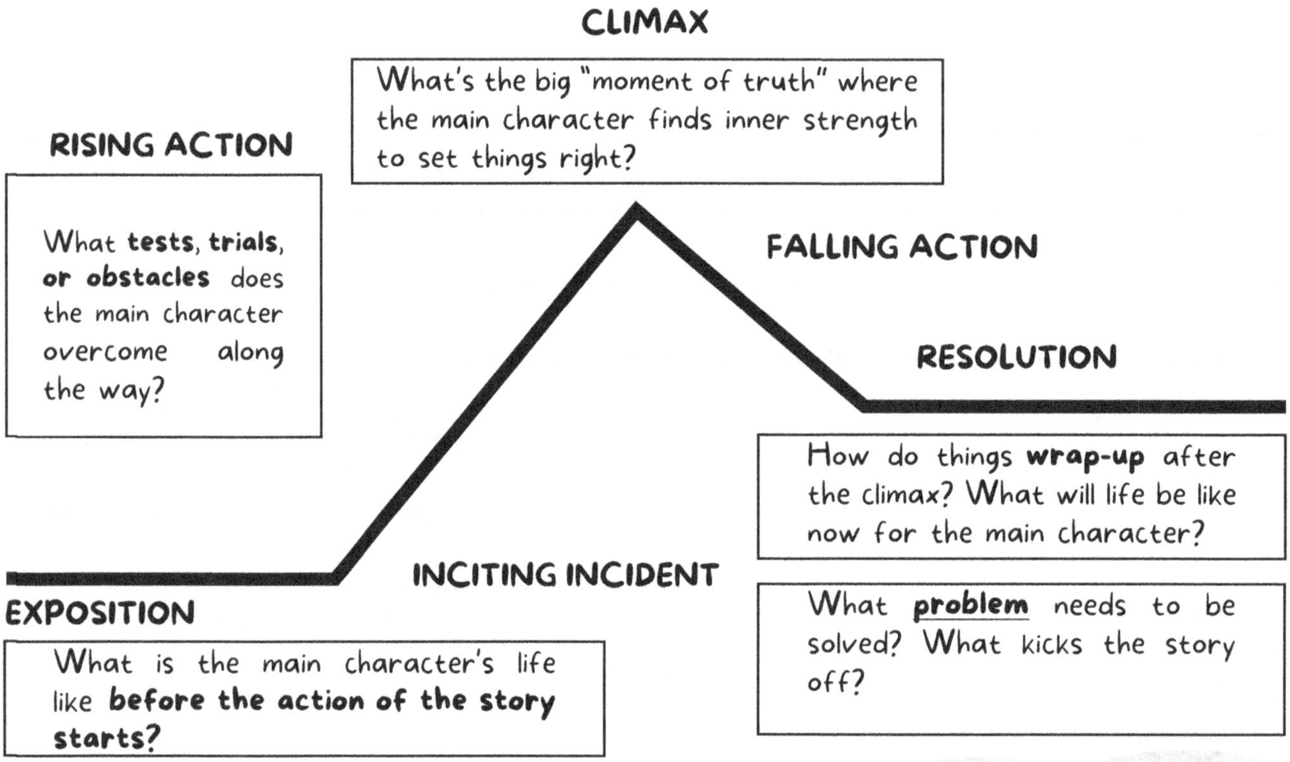

FITNESS

Please be aware of your environment and be safe at all times. If you cannot do an exercise, just try your best.

Repeat these exercises 4 ROUNDS

1 - Abs: 20 times

2 - Lunges: 8 times to each leg.
Note: Use your body weight or books as weight to do leg lunges.

3 - Plank: 20 sec.

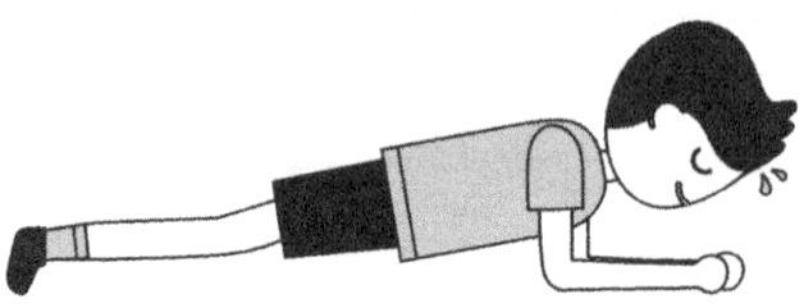

4 - Run: 50m
Note: Run 25 meters to one side and 25 meters back to the starting position.

READING PASSAGE

From "Pygmalion"

By Bernard Shaw

(Continued from Day 1's Passage)

MRS. PEARCE [hesitating, evidently perplexed] A young woman wants to see you, sir.

HIGGINS. A young woman! What does she want?

MRS. PEARCE. Well, sir, she says you'll be glad to see her when you know what she's come about. She's quite a common girl, sir. Very common indeed. I should have sent her away, only I thought perhaps you wanted her to talk into your machines. I hope I've not done wrong; but really you see such strange people sometimes - you'll excuse me, I'm sure, sir -

HIGGINS. Oh, that's all right, Mrs. Pearce. Has she an interesting accent?

MRS. PEARCE. Oh, something dreadful, sir, really. I don't know how you can take an interest in it.

HIGGINS [to Pickering] Let's have her up. Show her up, Mrs. Pearce [he rushes across to his working table and picks out a cylinder to use on the phonograph].

MRS. PEARCE [only half resigned to it] Very well, sir. It's for you to say. [She goes downstairs].

HIGGINS. This is rather a bit of luck. I'll show you how I make records. We'll set her talking; and I'll take it down first in Bell's visible Speech; then in broad Romic; and then we'll get her on the phonograph so that you can turn her on as often as you like with the written transcript before you.

MRS. PEARCE [returning] This is the young woman, sir.

The flower girl enters in state. She has a hat with three ostrich feathers, orange, sky-blue, and red. She has a nearly clean apron, and the shoddy coat has been tidied a little.

THE FLOWER GIRL. Did you tell him I come in a taxi?

MRS. PEARCE. Nonsense, girl! what do you think a gentleman like Mr. Higgins cares what you came in?

THE FLOWER GIRL. Oh, we are proud! He ain't above giving lessons, not him: I heard him say so. Well, I ain't come here to ask for any compliment; and if my money's not good enough I can go elsewhere.

HIGGINS. Good enough for what?

OVERVIEW OF ENGLISH CONCEPTS
READING PASSAGE

THE FLOWER GIRL. Good enough for ye - oo. Now you know, don't you? I'm come to have lessons, I am. And to pay for em too: make no mistake.

HIGGINS [stupent] WELL!!! [Recovering his breath with a gasp] What do you expect me to say to you?

THE FLOWER GIRL. Well, if you was a gentleman, you might ask me to sit down, I think. Don't I tell you I'm bringing you business?

HIGGINS. Pickering: shall we ask this baggage to sit down or shall we throw her out of the window?

THE FLOWER GIRL [running away in terror to the piano, where she turns at bay] Ah-ah-ah-ow-ow-ow-oo! [Wounded and whimpering] I won't be called a baggage when I've offered to pay like any lady.

Motionless, the two men stare at her from the other side of the room, amazed.

PICKERING [gently] What is it you want, my girl?

THE FLOWER GIRL. I want to be a lady in a flower shop stead of selling at the corner of Tottenham Court Road. But they won't take me unless I can talk more genteel. He said he could teach me. Well, here I am ready to pay him - not asking any favor - and he treats me as if I was dirt.

1. Based on the last two passages, what kind of work does it seem like **Higgins** does?

____

____

____

2. Why is Mrs. Pierce very **apprehensive** at the beginning of the scene?

____

____

____

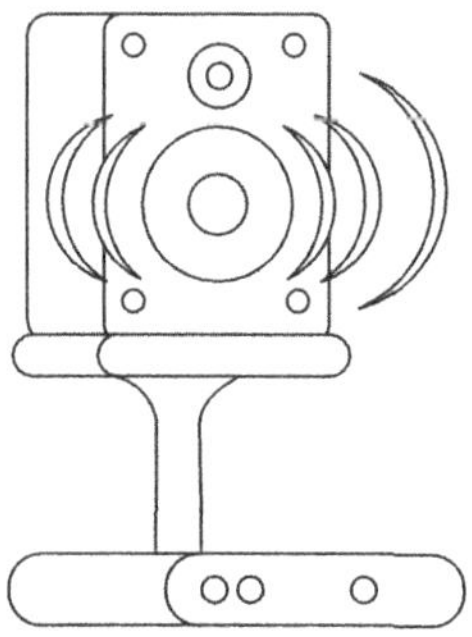

3. Why does the Flower Girl want Higgins to know she came in a taxi?

   A. To show that she is in a rush
   B. Because she wants him to pay for her return trip
   C. To show she has money
   D. Because she thinks he will be impressed

4. What does the Flower Girl want from Higgins?

   A. To give her money so she can get training for a better job
   B. To help her improve her strong accent so she can get a better job
   C. To help her improve her strong accent so people stop making fun of her
   D. To give her money so she can move to another town

5. Do you think the Flower Girl will be good at learning to speak more "genteel"? What about her character or personality makes you think so?

____

____

____

____

____

____

Creating Your Own

Directions: Last time, you took one of your **favorite stories** and used the plotline to break down and describe its structure. Today, you'll be **planning your own story** using a plotline. Brainstorm a story of your own and then fill out the boxes below to describe what would happen in it. (You may want to do some initial planning on scratch paper first.) Focus on having fun, thinking big, and providing your reader with action that rises toward a climax.

TEXT TITLE: ______________________________

CLIMAX

What's the big "moment of truth" where the main character finds inner strength to set things right?

RISING ACTION

What **tests, trials, or obstacles** does the main character overcome along the way?

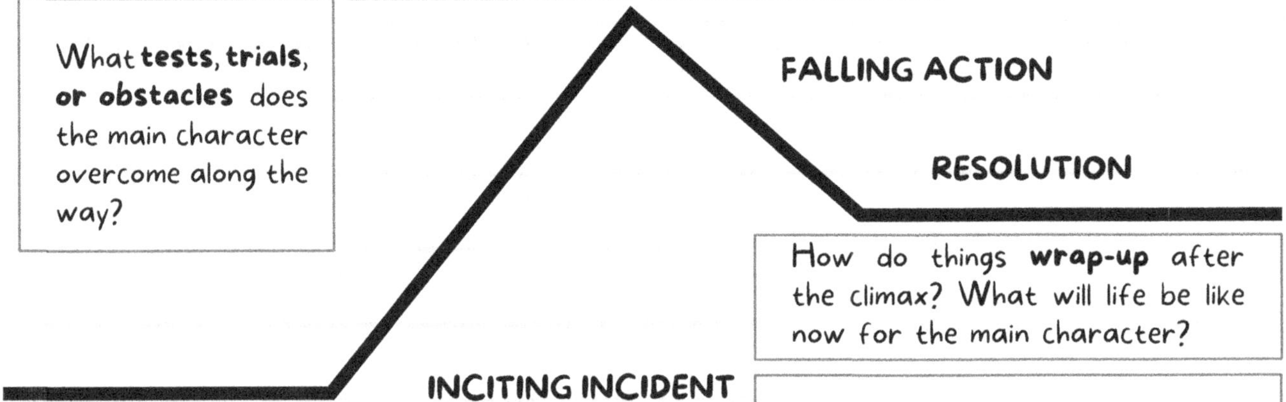

FALLING ACTION

RESOLUTION

How do things **wrap-up** after the climax? What will life be like now for the main character?

INCITING INCIDENT

What **problem** needs to be solved? What kicks the story off?

EXPOSITION

What is the main character's life like **before the action of the story starts?**

FITNESS

Please be aware of your environment and be safe at all times. If you cannot do an exercise, just try your best.

Repeat these exercises **4 ROUNDS**

1 - Squats: 20 times.
Note: imagine you are trying to sit on a chair.

2 - Side Bending: 15 times to each side. Note: try to touch your feet.

3 - Tree Pose: Stay as long as possible.
Note: do the same with the other leg.

WEEK 5 DAY 3

MATH

Expressions and Equations

1. $11y + 20y$

   A. $21y$
   B. $11y$
   C. $9y$
   D. $31y$

2. $14n - 8n$

   A. 6
   B. n
   C. $6n$
   D. $22n$

3. $3k - 19k$

   A. $16k$
   B. $-16k$
   C. $22k$
   D. $-22k$

4. $-z + 5z$

   A. $4z$
   B. $6z$
   C. $-4z$
   D. $-6z$

5. $8x^2 - 13x^2$

   A. $-21x^2$
   B. $21x^2$
   C. $-5x^2$
   D. $5x^2$

6. $5d - 4 + 3d - 14$

   A. $-8d + 18$
   B. -18
   C. $8d$
   D. $8d - 18$

7. $5 - 2q + 2r - 10$

   A. $-5 - 2q + 2r$
   B. $-5 + 2q + 2r$
   C. $2r - 2q + 5$
   D. $2q - 2r - 5$

8. $6 - x + y - 4y$

   A. $6 - 6y$
   B. $10xy$
   C. $6 - x - 3y$
   D. $6 + x + 3y$

9. $5n + 8 - 4n + m$

   A. $n - m - 8$
   B. $n + m + 8$
   C. $-n + m - 8$
   D. $n + m - 8$

10. $4n^2 - 11n + 4n + 2 - 9$

   A. $11n^3 - 7$
   B. $11n^3 + 7$
   C. $4n^2 + 7n + 7$
   D. $4n^2 - 7n - 7$

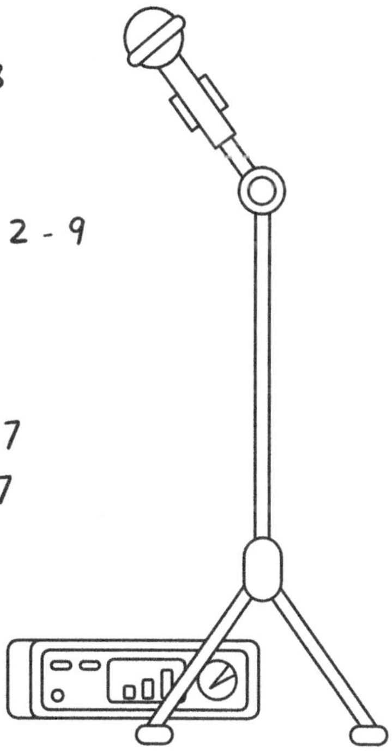

MATH

Expressions and Equations

11. $10(4x + 3)$

- **A.** $40x + 30$
- **B.** $40x + 3$
- **C.** $4x - 10$
- **D.** $4x-30$

12. $14(3m - n)$

- **A.** $3m - 14n$
- **B.** $42m - n$
- **C.** $42m - 14n$
- **D.** $42m + 14n$

13. $(2b + 4h)4$

- **A.** $8b + 16h$
- **B.** $8b + 16$
- **C.** $8b - 16h$
- **D.** $8b - 16$

14. $\frac{1}{2}\left(14a - \frac{1}{2}b\right)$

- **A.** $7a - \frac{1}{4}b$
- **B.** $7a-b$
- **C.** $7a - \frac{1}{2}b$
- **D.** $7a + \frac{1}{4}b$

15. $-h(h - 4)$

- **A.** $-4h - h^2$
- **B.** $4h - h$
- **C.** $-h^2 + 4h$
- **D.** $h^2 - 4h$

FITNESS

Please be aware of your environment and be safe at all times. If you cannot do an exercise, just try your best.

Repeat these exercises **4 ROUNDS**

1 - Bend forward: 15 times.
Note: try to touch your feet. Make sure to keep your back straight and if needed you can bend your knees.

2 - Lunges: 12 times to each leg.
Note: Use your body weight or books as weight to do leg lunges.

3 - Plank: 20 sec.

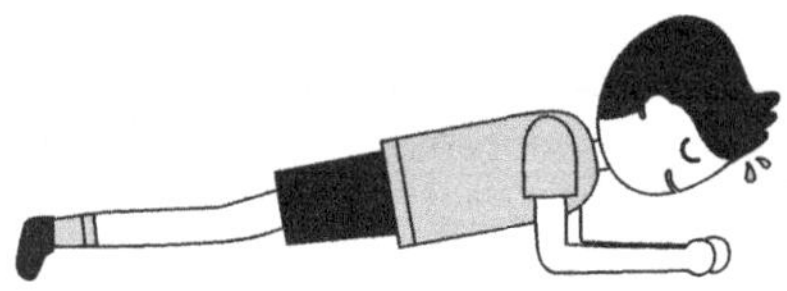

4 - Abs: 20 times

WEEK 5
DAY 4

MATH

Expressions and Equations

1. Which expression is represented by: y minus 15

   A. $y - 15$
   B. $y + 15$
   C. $15 - y$
   D. $15 + y$

2. Which expression is represented by: the product of 13 and h

   A. $13h$
   B. $13 + h$
   C. $13 - h$
   D. $\frac{h}{13}$

3. Which expression is represented by: the quotient of 11 and k

   A. $11k$
   B. $\frac{11}{k}$
   C. $11 - k$
   D. $k - 11$

4. Which expression is represented by: the sum of c and 8

   A. $\frac{c}{8}$
   B. $c - 8$
   C. $c + 8$
   D. $8c$

5. Which expression is represented by: a number r divided by 6

   A. $6 - r$
   B. $6r$
   C. $\frac{6}{r}$
   D. $\frac{r}{6}$

6. Which expression is represented by: the sum of 17 and twice a number g

   A. $2 - 17g$
   B. $17g - 2$
   C. $2g - 17$
   D. $2g + 17$

7. Which expression is represented by: 5.8 more than the product of 3 and u

   A. $3u - 5.8$
   B. $3u + 5.8$
   C. $5.8 - 3u$
   D. $3 - 5.8u$

8. Which expression is represented by: 10 minus the quotient of d and 5

   A. $\frac{d}{5} - d$
   B. $10 + \frac{d}{5}$
   C. $10 - \frac{d}{5}$
   D. $\frac{d}{5} - 10$

WEEK 5 DAY 4 MATH

Expressions and Equations

9. Which expression is represented by: 12 plus the quotient of 60 and r

A. $\frac{r}{60} + 12$

B. $12 + \frac{60}{r}$

C. $\frac{60}{r} - 12$

D. $12 - \frac{r}{60}$

10. Which expression is represented by: the product of $\frac{1}{4}$ and the sum of e and 11

A. $\frac{1}{4}(11 + e)$

B. $11\left(\frac{1}{4}e\right)$

C. $\frac{1}{4} + 11 + e$

D. $\frac{1}{4}(11 - e)$

11. Write a word phrase for the following expression: $48 + 4n$

Answer: ______________________

12. Write a word phrase for the following expression: $\frac{z}{4} + 5$

Answer: ______________________

13. Write a word phrase for the following expression: $9w - 2$

Answer: ______________________

14. Write a word phrase for the following expression: $14 + 0.7s$

Answer: ______________________

15. Write a word phrase for the following expression: $3(d - 4)$

Answer: ______________________

FITNESS

Please be aware of your environment and be safe at all times. If you cannot do an exercise, just try your best.

Repeat these exercises **4 ROUNDS**

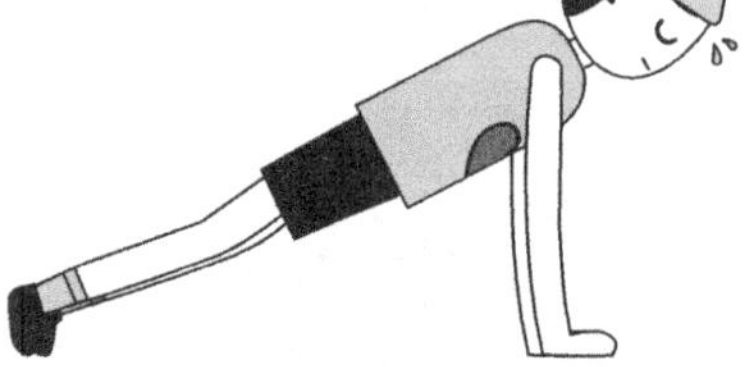

2 - Chair: 20 sec.
Note: sit on an imaginary chair, keep your back straight.

1 - High Plank: 25 sec.

3 - Waist Hooping: 20 times. Note: if you do not have a hoop, pretend you have an imaginary hoop and rotate your hips 10 times.

4 - Abs: 25 times

WEEK 5 DAY 5

MATH

Expressions and Equations

1. $\sqrt{36}$

A. 6
B. 5
C. 4
D. 3

2. $\sqrt{900}$

A. 45
B. 300
C. 20
D. 30

3. $\sqrt{\frac{1}{9}}$

A. $\frac{1}{3}$
B. $\frac{1}{2}$
C. $\frac{3}{1}$
D. $\frac{1}{8}$

4. $\sqrt{1.96}$

A. 1.4
B. 1.3
C. 1.2
D. 1.1

5. $\sqrt{\frac{144}{16}}$

A. 2
B. 3
C. 4
D. 5

6. Estimate and round to the nearest integer. $\sqrt{18}$

A. 5
B. 2
C. 3
D. 4

7. Estimate and round to the nearest integer. $\sqrt{42}$

A. 4
B. 5
C. 6
D. 7

8. Estimate and round to the nearest integer. $\sqrt{260}$

A. 14
B. 15
C. 16
D. 17

9. Estimate and round to the nearest integer. $\sqrt{62}$

A. 8
B. 7
C. 6
D. 5

10. Estimate and round to the nearest integer. $\sqrt{315}$

A. 17
B. 18
C. 19
D. 20

11. Rewrite as a rational number in the form $\frac{a}{b}$: 412

A. $\frac{412}{10}$
B. $\frac{10}{412}$
C. $\frac{1}{412}$
D. $\frac{412}{1}$

WEEK 5
DAY 5

MATH

Expressions and Equations

12. Rewrite as a rational number in the form $\frac{a}{b}$: 0.72

A. $\frac{72}{10}$

B. $\frac{7}{10}$

C. $\frac{72}{100}$

D. $\frac{100}{72}$

13. Rewrite as a rational number in the form $\frac{a}{b}$: 3.28

A. $\frac{328}{100}$

B. $\frac{32}{100}$

C. $\frac{300}{1}$

D. $\frac{32}{8}$

14. Rewrite as a rational number in the form $\frac{a}{b}$: 5.4

A. $\frac{54}{10}$

B. $\frac{54}{100}$

C. $\frac{5}{4}$

D. $\frac{5}{10}$

15. Rewrite as a rational number in the form $\frac{a}{b}$: 8.92

A. $\frac{89}{2}$

B. $\frac{89}{10}$

C. $\frac{89}{100}$

D. $\frac{892}{100}$

YOGA

Please be aware of your environment and be safe at all times. If you cannot do an exercise, just try your best.

1 - Down Dog: 30 sec.

2 - Bend Down: 30 sec.

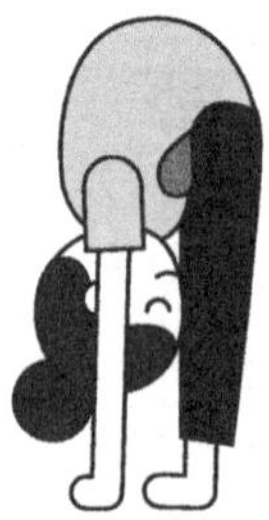

3 - Chair: 30 sec.

4 - Child Pose: 30 sec.

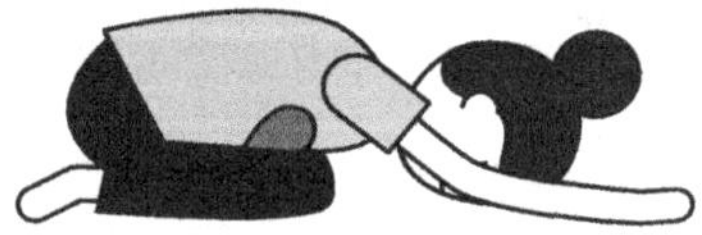

5 - Shavasana: as long as you can. Note: think of happy moments and relax your mind.

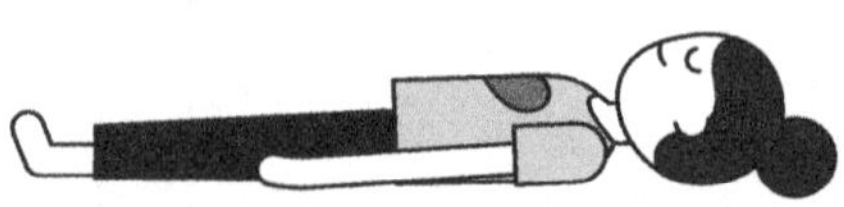

EXPERIMENT

Harnessing Thermal Energy

During Week 4's experiment, you saw how increasing ***temperature*** *allows* ***particles*** *to* ***move*** *much more energetically. This week, we'll do an activity to show how some of that* ***energy*** *can be harnessed by people for use. For example, steam engines have been used for centuries to turn heat into usable energy. In this experiment, you'll create a miniature steam engine that uses heat to accelerate particles, create motion, and generate force.*

Materials:

- 1 unopened soda can
- 1 thumbtack
- A sink with running water
- A pillar candle
- String
- Scissors
- A desk lamp with an arm
- Paper towels

WEEK 5 DAY 6 EXPERIMENT

Procedure:

1. Place your unopened soda can in an empty sink on its side.

2. Using your thumbtack, poke a hole in the side of the can (near the middle). Shake the can back and forth inside the sink to build up pressure and shoot all the soda out of the can (be sure to clean up anything that sprays outside of the sink using paper towels).

3. Once the can is empty, rotate it and create a second hole directly across from your first hole on the other side of the can.

4. Rinse the can out thoroughly without opening the tab at the top. Use the small holes you've created to spray water inside the can and dump it out again.

5. Once the can is thoroughly rinsed, use the hole to allow about 2 teaspoons of water into the bottom of the can.

6. Set up your desk lamp on a counter or table where you can extend its arm. Place the pillar candle directly underneath the arm of the lamp. Using your scissors, cut a piece of string that is about as long as the distance from the arm of the lamp down to the surface of the table or counter.

7. Tie one end of the string you've just cut to the arm of the lamp. Tie the other end to the pull tab on the top of the soda can (being careful not to puncture the seal). Be sure the can is well-secured.

8. Arrange the lamp so it is dangling the can directly above the pillar candle.

9. Light the pillar candle **and step back.** (You don't want to be too close for this!)

10. Observe the can as the candle warms and ultimately boils the water inside of it. When steam begins to appear from the holes, observe the **motion** of the can and brainstorm different uses for a steam engine like this.

11. Wait patiently for the steam to subside. Then, blow out the candle, answer the questions below, and clean up your materials.

WEEK 5 DAY 6 EXPERIMENT

Follow-Up Questions:

1. How were the results of this experiment/activity similar to, or different from, the Week 4 activity?

____

____

____

2. Name two different ways using **heat to create motion** in this way might be useful to people:

____

____

____

YOGA

Please be aware of your environment and be safe at all times. If you cannot do an exercise, just try your best.

2 - Down Dog: 30 sec.

3 - Stretching: Stay as long as possible.
Note: do on one leg then on another.

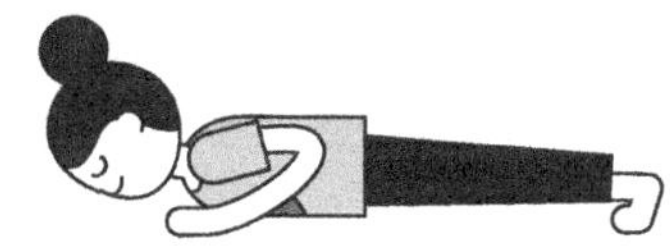

4 - Lower Plank: 20 sec.
Note: Keep your back straight and body tight.

1 - Tree Pose: Stay as long as possible.
Note: do on one leg then on another.

6 - Shavasana: 5 min.
Note: this pose is very important and provides you with long term benefits. Try not to skip this. Close your eyes and imagine who you want to be and what your goals are! Always think happy thoughts.

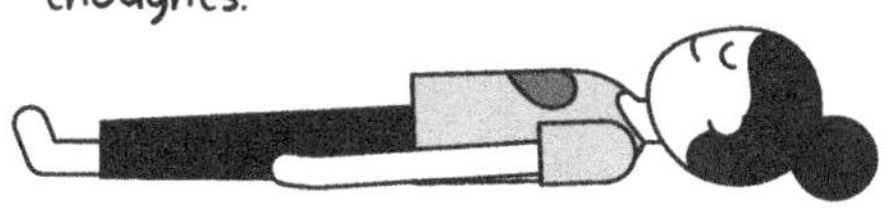

5 -Book Pose: 20 sec.
Note: Keep your core tight. Legs should be across from your eyes.

WEEK 5 DAY 7 LOGIC

Task:

There are 18 images below. Pair one of the images with another image to form a triangle. There are a total of 9 triangles. List the corresponding pairs.

WEEK 6

ARGOPREP

WEEK 6 DAY 1 — OVERVIEW OF ENGLISH CONCEPTS

The Relationships Between Scenes & Plot Points

Now that you're familiar with the basic structure that most stories follow, it's time to dig deep and get a little more specific. Last week, we said that most narratives begin with a period of **exposition** before an **inciting incident** kicks off the **rising action** of the plot, in which things grow in importance, intensity, and (hopefully) interest level. This week, we'll zero in on that stretch of the story to think about what order things happen in and why they should (or need to) happen that way.

As we said last week, during a story, characters face **increasingly tough** challenges, which they have to overcome by **digging deep within themselves, learning new things, forging important partnerships, or finding the right tools**. Making sure those events build in a natural, exciting way toward the **climax** of the narrative is one of the author's most important jobs. This week, we'll approach this topic from both the author and the reader's perspective.

Thinking About the Progression of the Plot as a Reader:

- It's a **really good idea** to keep notes as you read. They **don't** have to be super detailed or follow any particular format, but one of the most important things you can do to set yourself up for success is just to keep a few notes about **plot, character development, and things that stuck out to you**.
    - For example, after you read a chapter in a novel, write a one or two sentence summary of what happened and write down any specific things that **happened** or what **the characters said** that seemed memorable. You can also make note of sentences or ideas from the book that you **like!**
- Even if you're not keeping notes, it's a great idea to check in with yourself after reading a section or chapter of a text and ask a few questions...
    - What just happened?
    - How is this connected to **what happened before it**?
    - What different things **could possibly happen now** based on what I just read?
- Pay attention to how the plot **builds** in intensity.
    - Each time the characters face a new challenge or encounter a new situation, think about how that situation is **similar to or different from** the challenges they've already faced and overcome. Ask yourself how that particular challenge or episode represents an **escalation** in intensity, suspense, or interest level over what's come before it.

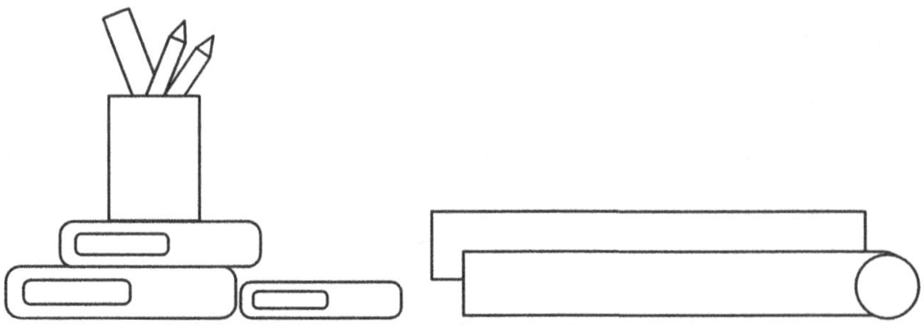

WEEK 6 DAY 1 OVERVIEW OF ENGLISH CONCEPTS

As a Writer...

- Don't let your narrative become **flat.**
  - Make sure that you don't **start the story** with your more interesting content, only to follow it without any **escalating or rising action/suspense/drama.**
- Always make sure things **make sense and are easy to follow.**
  - Read your work multiple times to make sure the plot and main ideas of your story are easy to understand.
  - Using a peer editor, friend, teacher, or family member as a second set of eyes in this regard is also extremely beneficial!

READING PASSAGE

From "Pygmalion"

By Bernard Shaw

(Continued from last week's passages)

THE FLOWER GIRL. I want to be a lady in a flower shop stead of selling at the corner of Tottenham Court Road. But they won't take me unless I can talk more genteel. He said he could teach me. Well, here I am ready to pay him — not asking any favor — and he treats me as if I was dirt.

MRS. PEARCE. How can you be such a foolish ignorant girl as to think you could afford to pay Mr. Higgins?

THE FLOWER GIRL. Why shouldn't I? I know what lessons cost as well as you do; and I'm ready to pay.

HIGGINS. How much?

THE FLOWER GIRL [coming back to him, triumphant] Now you're talking

HIGGINS [peremptorily] Sit down.

MRS. PEARCE [severely] Sit down, girl. Do as you're told. [She places the stray chair near the hearthrug between Higgins and Pickering, and stands behind it waiting for the girl to sit down].

THE FLOWER GIRL. Ah-ah-ah-ow-ow-oo! [She stands, half rebellious, half bewildered].

PICKERING [very courteous] Won't you sit down?

LIZA [coyly] Don't mind if I do. [She sits down. Pickering returns to the hearthrug].

HIGGINS. What's your name?

THE FLOWER GIRL. Liza Doolittle.

HIGGINS. How much do you propose to pay me for the lessons?

LIZA. Oh, I know what's right. A lady friend of mine gets French lessons for eighteenpence an hour from a real French gentleman. Well, you wouldn't have the face to ask me the same for teaching me my own language as you would for French; so I won't give more than a shilling. Take it or leave it.

WEEK 6 DAY 1

OVERVIEW OF ENGLISH CONCEPTS
READING PASSAGE

HIGGINS [walking up and down the room, rattling his keys and his cash in his pockets] You know, Pickering, if you consider a shilling, not as a simple shilling, but as a percentage of this girl's income, it works out as fully equivalent to sixty or seventy guineas from a millionaire.

PICKERING. How so?

HIGGINS. Figure it out. A millionaire has about 150 pounds a day. She earns about half-a-crown.

LIZA [haughtily] Who told you I only-

HIGGINS [continuing] She offers me two-fifths of her day's income for a lesson. Two-fifths of a millionaire's income for a day would be somewhere about 60 pounds. It's handsome. By George, it's enormous! it's the biggest offer I ever had.

LIZA [rising, terrified] Sixty pounds! What are you talking about? I never offered you sixty pounds. Where would I get-

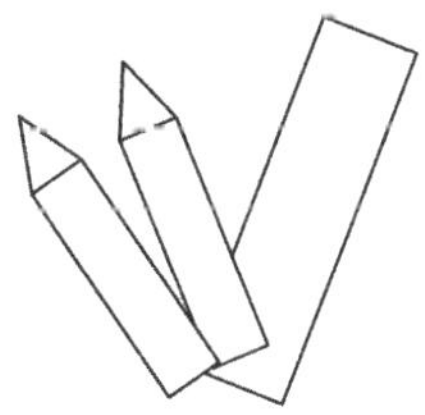

HIGGINS. Hold your tongue.

LIZA [weeping] But I ain't got sixty pounds. Oh-

MRS. PEARCE. Don't cry, you silly girl. Sit down. Nobody is going to touch your money.

1. Which details from this scene show that **Liza** currently does not command a great deal of <u>respect</u>?

____

____

____

2. Why does Liza cry at the end of the scene?

____

____

____

3. How did **Liza** decide how much to offer Higgins?

A. She saw how much he charged in the newspaper.
B. She offered what she thought was fair.
C. She offers him everything she has.
D. She based the price on how much French lessons cost.

4. Why is Higgins **impressed** by Liza's proposed payment?

A. He is impressed by how brave she is for coming to talk to him.
B. She is offering him more than wealthy people do.
C. She is offering him a higher percentage of her income than almost anybody else does.
D. She has developed a clear plan to pay him over time, even though she can't pay him now.

5. Do you think Higgins will coach Liza at this point? **Why** or **why not**?

Writing Sequential Scenes that Make Sense

Directions: Look back over the **plotline that you created for your own story on Day 2 of last week** and see what kind of tests or adventures you said took place in the rising action of the story. Today, you'll be thinking about how you could craft **two** scenes or moments from your story in a way that builds naturally and logically. Answer the questions below to guide your thinking.

1. What are **two** important tests or challenges that your main character will overcome in the **Rising Action** portion of your story? (These could be mental puzzles they have to solve, enemies they have to defeat, sketchy situations they have to escape, etc.)

   a. ______________________________

   ____

   b. ______________________________

   ____

2. What kind of **skills** or **knowledge** will your main character need to solve the first test/challenge/obstacle?

____

____

____

3. What important **lesson**, **skill**, **treasure** or **way of looking at things** will your main character gain from overcoming/solving the first challenge?

____

____

OVERVIEW OF ENGLISH CONCEPTS
ACTIVITIES

4. How can the **lessons**, **skills**, **treasures**, **and ways of looking at things** the main character gained during the first challenge become an important part of the second challenge? (Or, you can think of it this way: How will you **connect** the second challenge to the first challenge to show the character has grown, changed, and gotten better or more capable?)

5. What important **lesson**, **skill**, **treasure,** or **way of looking at things** will your main character gain from overcoming/solving the second challenge?

6. How might **those skills, treasures, etc.** be relevant to future scenes in the text (especially the **climax**)?

FITNESS

Please be aware of your environment and be safe at all times. If you cannot do an exercise, just try your best.

Repeat these exercises **4 ROUNDS**

2 - Lunges: 8 times to each leg.
Note: Use your body weight or books as weight to do leg lunges.

1 - Abs: 20 times

3 - Plank: 20 sec.

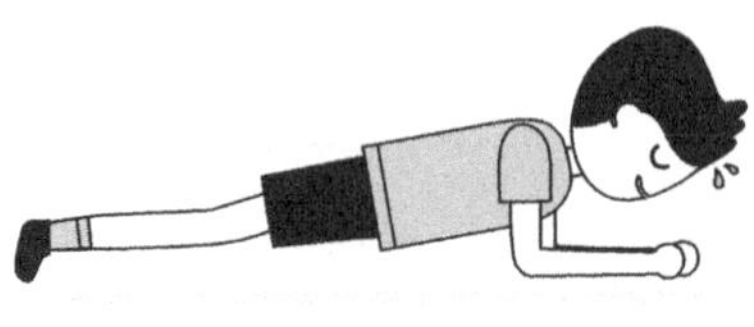

4 - Run: 50m
Note: Run **25** meters to one side and **25** meters back to the starting position.

READING PASSAGE

From "Pygmalion"

By Bernard Shaw

HIGGINS [rising hastily and running to Mrs. Higgins] Here she is, mother. [He stands on tiptoe and makes signs over his mother's head to Eliza to indicate to her which lady is her hostess].

Eliza, who is exquisitely dressed, produces an impression of such remarkable distinction and beauty as she enters that they all rise, quite flustered. Guided by Higgins's signals, she comes to Mrs. Higgins with studied grace.

LIZA [speaking with pedantic correctness of pronunciation and great beauty of tone] How do you do, Mrs. Higgins? [She gasps slightly in making sure of the H in Higgins, but is quite successful]. Mr. Higgins told me I might come.

MRS. HIGGINS [cordially] Quite right: I'm very glad indeed to see you.

PICKERING. How do you do, Miss Doolittle?

LIZA [shaking hands with him] Colonel Pickering, is it not?

MRS. EYNSFORD HILL. I feel sure we have met before, Miss Doolittle. I remember your eyes.

LIZA. How do you do? [She sits down on the ottoman gracefully in the place just left vacant by Higgins].

MRS. EYNSFORD HILL [introducing] My daughter Clara.

LIZA. How do you do?

CLARA [impulsively] How do you do? [She sits down on the ottoman beside Eliza, devouring her with her eyes].

FREDDY [coming to their side of the ottoman] I've certainly had the pleasure.

MRS. EYNSFORD HILL [introducing] My son Freddy.

LIZA. How do you do?

Freddy bows and sits down in the Elizabethan chair, infatuated.

HIGGINS [suddenly] By George, yes: it all comes back to me! [They stare at him]. Covent Garden!

READING PASSAGE

MRS. HIGGINS. Henry, please! [He is about to sit on the edge of the table]. Don't sit on my writing-table: you'll break it.

HIGGINS [sulkily] Sorry.

A long and painful pause ensues.

MRS. HIGGINS [at last, conversationally] Will it rain, do you think?

LIZA. The shallow depression in the west of these islands is likely to move slowly in an easterly direction. There are no indications of any great change in the barometrical situation.

FREDDY. Ha! ha! how awfully funny!

LIZA. What is wrong with that, young man? I bet I got it right.

FREDDY. Killing!

MRS. EYNSFORD HILL. I'm sure I hope it won't turn cold. There's so much influenza about. It runs right through our whole family regularly every spring.

LIZA [darkly] My aunt died of influenza: so they said.

MRS. EYNSFORD HILL [clicks her tongue sympathetically]!!!

LIZA [in the same tragic tone] But it's my belief they done the old woman in.

MRS. HIGGINS [puzzled] Done her in?

1. What other names do we know **Mrs. Eynsford Hill** and **Clara** by from previous scenes?

____

____

2. How is **Eliza/Liza/The Flower Girl** different in this scene than she has been in previous scenes?

____

____

____

____

OVERVIEW OF ENGLISH CONCEPTS
READING PASSAGE

3. Why does Freddy **laugh** at Liza?

   A. Because he thinks she is stupid
   B. Because she tells a joke
   C. Because all the other characters are laughing as well
   D. Because her attempts to speak politely are overly technical

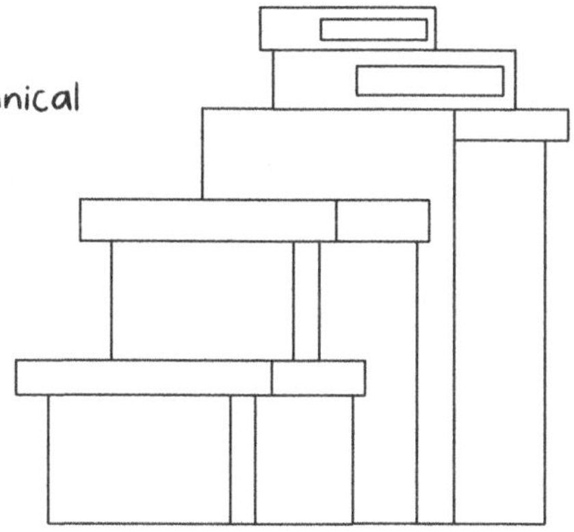

4. What does **Liza** believe happened to her aunt?

   A. She believes her aunt died of the flu.
   B. She believes her aunt faked her death.
   C. She believes her aunt was murdered.
   D. She believes Mrs. Eynsford Hill is her aunt.

5. Which lines or ideas in this passage show that, while Liza has learned a lot, she **still stands out** when communicating with rich, upper-class people?

___

___

___

Writing Scenes in a Logical, Sequential Manner

Directions: On Day 1, you planned how you could craft **two scenes or moments** from the rising action of the story you first brainstormed last week. Now, it's time to **actually draft those scenes**. Use your notes from Day 2 of last week and Day 1 of this week to write two short scenes below. The second scene should logically follow the first scene and include **skills, knowledge, or objects** that the main character gained in the first scene. Each scene should be **at least** six sentences.

OVERVIEW OF ENGLISH CONCEPTS

ACTIVITIES

FITNESS

Please be aware of your environment and be safe at all times. If you cannot do an exercise, just try your best.

Repeat these exercises **4 ROUNDS**

1 - Squats: 20 times.
Note: imagine you are trying to sit on a chair.

2 - Side Bending: 15 times to each side. Note: try to touch your feet.

3 - Tree Pose:
Stay as long as possible.
Note: do the same with the other leg.

WEEK 6 DAY 3 MATH

Expressions and Equations

1. $6b - 14 = 10$

   A. $b = 2$
   B. $b = 3$
   C. $b = 4$
   D. $b = 5$

2. $8a + 11 = 75$

   A. $a = 8$
   B. $a = 9$
   C. $a = 10$
   D. $a = 11$

3. $15 - 2n = 3$

   A. $n = 6$
   B. $n = 5$
   C. $n = 4$
   D. $n = 3$

4. $14 - 6x = 26$

   A. $x = -6$
   B. $x = -2$
   C. $x = 2$
   D. $x = 6$

5. $6r - 11 = 19$

   A. $r = -6$
   B. $r = 6$
   C. $r = -5$
   D. $r = 5$

6. $11 - 2y = -31$

   A. $y = 9$
   B. $y = 10$
   C. $y = 21$
   D. $y = -21$

7. $7g - 11 = 52$

   A. $g = -10$
   B. $g = 9$
   C. $g = -9$
   D. $g = 8$

8. $-2w + 8 = 24$

   A. $w = -8$
   B. $w = 8$
   C. $w = 3$
   D. $w = -3$

9. $10x + 7 = -63$

   A. $x = -7$
   B. $x = -8$
   C. $x = -9$
   D. $x = -10$

10. $6b - 10 = 56$

    A. $b = 11$
    B. $b = 10$
    C. $b = 9$
    D. $b = 8$

WEEK 6 DAY 3 MATH

11. $12j - 100 = 44$

A. $j = -10$
B. $j = 10$
C. $j = 12$
D. $j = -12$

12. $14 - 2t = 22$

A. $t = 4$
B. $t = -4$
C. $t = 2$
D. $t = -2$

13. $\frac{10q}{7} = 20$

A. $q = -14$
B. $q = 10$
C. $q = 12$
D. $q = 14$

14. $\frac{4i}{2} = 10$

A. $i = 5$
B. $i = 4$
C. $i = 3$
D. $i = 2$

15. $-2w + 14 = 38$

A. $w = -12$
B. $w = 12$
C. $w = 10$
D. $w = -10$

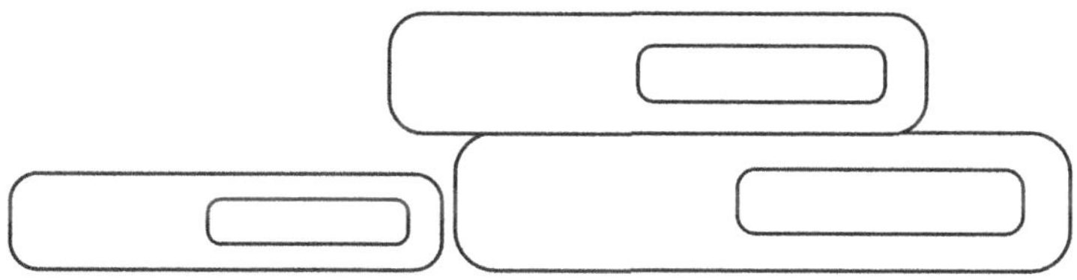

FITNESS

Please be aware of your environment and be safe at all times. If you cannot do an exercise, just try your best.

Repeat these exercises **4 ROUNDS**

1 - Bend forward: 15 times.
Note: try to touch your feet. Make sure to keep your back straight and if needed you can bend your knees.

2 - Lunges: 12 times to each leg.
Note: Use your body weight or books as weight to do leg lunges.

3 - Plank: 20 sec.

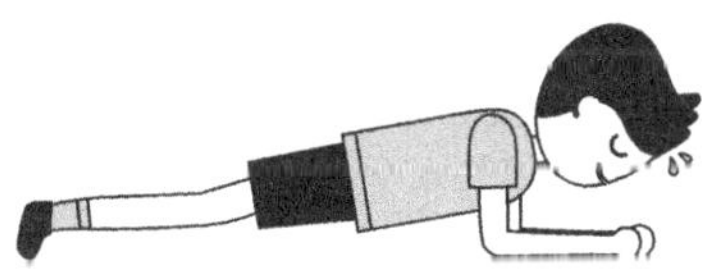

4 - Abs: 20 times

WEEK 6 DAY 4 MATH

Expressions and Equations

1. Which number is a solution of $4x - 2 > 5$

A. 2
B. 1
C. $\frac{1}{2}$
D. $\frac{1}{4}$

2. Which number is a solution of $12b + 10 \geq 16$

A. $\frac{1}{16}$ **C.** $\frac{1}{2}$
B. $\frac{1}{8}$ **D.** $\frac{1}{4}$

3. Which number is a solution of $4x + 2 < -6$

A. 1 **C.** -2
B. -1 **D.** -3

4. Which number is a solution of $c(c - 2) > 45$

A. 8 **C.** 6
B. 7 **D.** 5

5. Which number is a solution of $3y - 3 < 2$

A. 2 **C.** 3
B. 1 **D.** 4

6. What is the inequality of the graph?

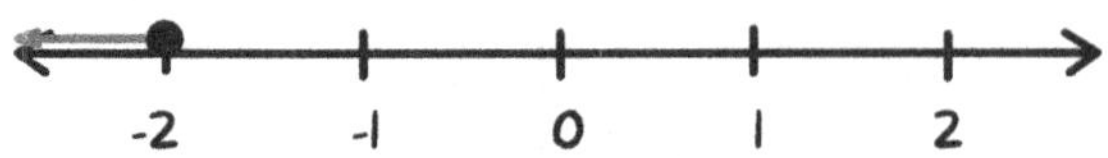

A. $x \geq -2$ **C.** $x \leq -2$
B. $x > 2$ **D.** $x > -2$

7. What is the inequality of the graph?

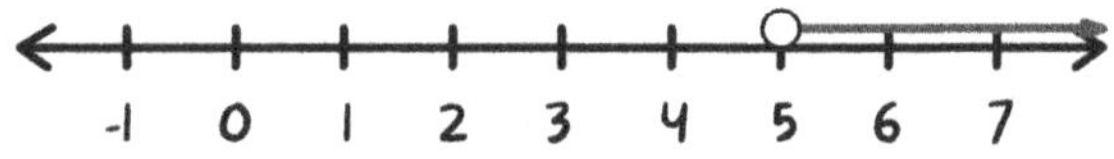

A. $x > 5$ **C.** $x > -5$
B. $x < 5$ **D.** $x < -5$

8. What is the inequality of the graph?

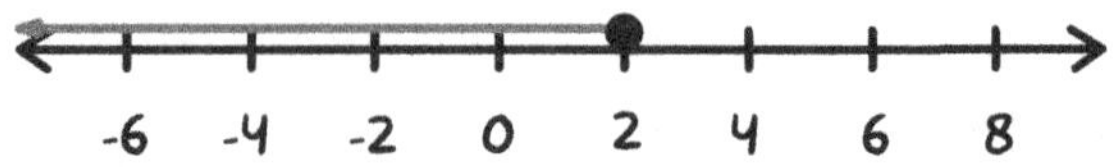

A. $y > -2$
B. $y \leq -2$
C. $y \geq 2$
D. $y \leq 2$

9. What is the inequality of the graph?

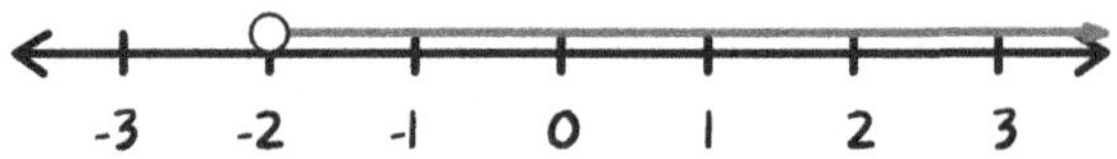

A. $b \geq 2$
B. $b > 2$
C. $b > -2$
D. $b \geq -2$

10. What is the inequality of the graph?

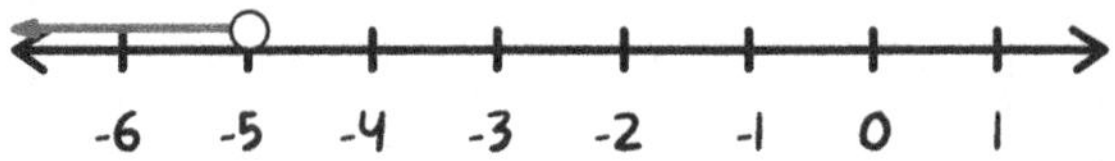

A. $n < -5$
B. $n < -6$
C. $n < -4$
D. $n < -3$

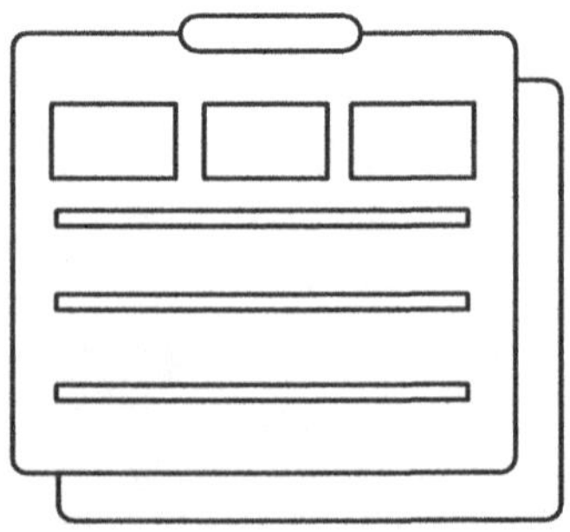

WEEK 6 DAY 4 MATH

11. Which inequality represents the following situation? A theater can seat no more than 88 people.

A. $s - 88 < 88$
B. $s + 88 > 88$
C. $s \leq 88$
D. $s > 88$

12. Which inequality represents the following situation? A person must be at least 35 to be President of the United States.

A. $P \geq 35$
B. $P \leq 35$
C. $P > 35$
D. $P < 35$

13. Which inequality represents the following situation? A light bulb should be no stronger than 66 watts.

A. $i \geq 66$
B. $i < 66$
C. $i > 66$
D. $i \leq 66$

14. Which inequality represents the following situation? At least 45 parents attended the meeting last week.

A. $45 > A + 45$
B. $45 < A + 45$
C. $A \geq 45$
D. $A \leq 45$

15. Which inequality represents the following situation? An attorney has earned at least 115,000 this year.

A. $E \geq 115000$
B. $E < 115000$
C. $E \leq 115000$
D. $E > 115$

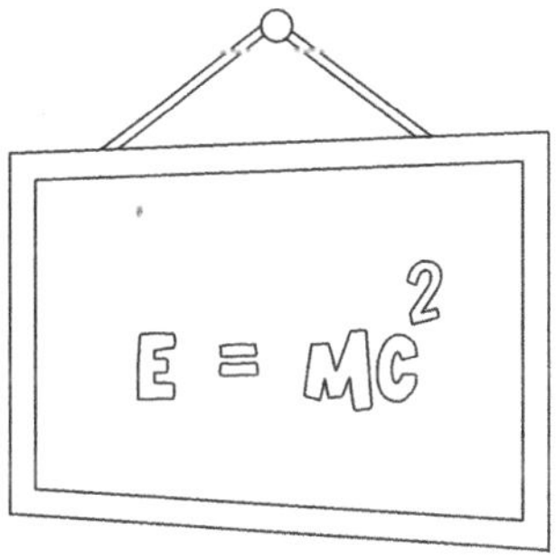

FITNESS

Please be aware of your environment and be safe at all times. If you cannot do an exercise, just try your best.

Repeat these exercises **4 ROUNDS**

1 - High Plank: 25 sec.

2 - Chair: 20 sec.
Note: sit on an imaginary chair, keep your back straight.

3 - Waist Hooping: 20 times. Note: if you do not have a hoop, pretend you have an imaginary hoop and rotate your hips 10 times.

4 - Abs: 25 times

MATH

Expressions and Equations

1. $m - 3.6 > 2.4$

   A. $m > 6$
   B. $m < 6$
   C. $m > -6$
   D. $m < -6$

2. $n + 5.1 < 3.6$

   A. $n > -1.5$
   B. $n > 1.5$
   C. $n < 1.5$
   D. $n < -1.5$

3. $4 \leq r - 3$

   A. $r \leq 7$
   B. $r \geq 7$
   C. $r > 7$
   D. $r < 7$

4. $15 \geq b + 3$

   A. $b \geq -12$
   B. $b \leq -12$
   C. $b \leq 12$
   D. $b \geq 12$

5. $\frac{x}{4} > 6$

   A. $x < -24$
   B. $x > -24$
   C. $x < 24$
   D. $x > 24$

6. $10k \leq 9$

   A. $k \geq \frac{10}{9}$
   B. $k \geq \frac{9}{10}$
   C. $k \leq \frac{9}{10}$
   D. $k \leq \frac{10}{9}$

7. $11 \geq b - 22$

   A. $b \leq 33$
   B. $b \leq -33$
   C. $b > 33$
   D. $b > -2$

8. $\frac{a}{5} < 7$

   A. $a < 35$
   B. $a > 35$
   C. $a < -35$
   D. $a > -35$

9. $-3d > 15$

   A. $d > -5$
   B. $d < -5$
   C. $d < 5$
   D. $d > 5$

10. $-7f < -56$

    A. $f < 8$
    B. $f < -8$
    C. $f > -8$
    D. $f > 8$

WEEK 6
DAY 5

MATH

11. $4n - 2 > 42$

A. $n > 11$
B. $n < 11$
C. $n > -11$
D. $n < -11$

12. $2 + 5p < 17$

A. $p > -3$
B. $p < -3$
C. $p < 3$
D. $p > 3$

13. $12 - 3b \leq 21$

A. $b \leq 3$
B. $b \geq 3$
C. $b \leq -3$
D. $b \geq -3$

14. $50 - 5c > 75$

A. $c < -5$
B. $c > -5$
C. $c < 5$
D. $c > 5$

15. $11 + 8j \geq 35$

A. $j > 3$
B. $j < 3$
C. $j \leq 3$
D. $j \geq 3$

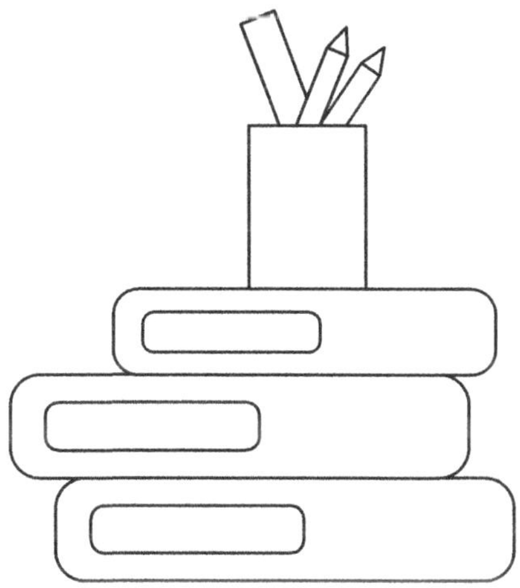

YOGA

Please be aware of your environment and be safe at all times. If you cannot do an exercise, just try your best.

1 - Down Dog: 30 sec.

2 - Bend Down: 30 sec.

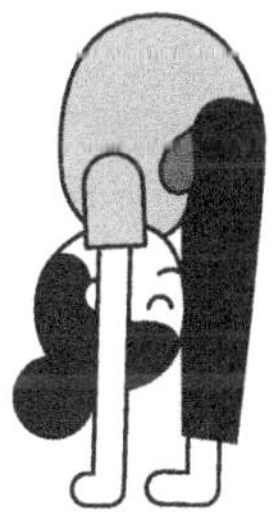

3 - Chair: 30 sec.

4 - Child Pose: 30 sec.

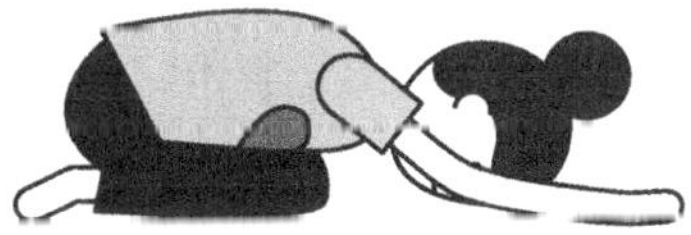

5 - Shavasana: as long as you can. Note: think of happy moments and relax your mind.

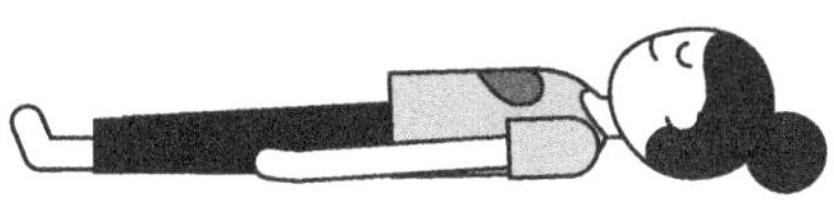

EXPERIMENT

Temperature & Density in Air Masses

*Over the last two weeks, we've looked at how **temperature** affects the **movement** of particles and can be used to create **heat energy**. Now it's time to think about how some of the heat exchanges we've looked at occur in nature. In the atmosphere above us, masses of air and water vapor of varying temperatures are constantly interacting with each other, creating our **weather**.*

In today's experiment, you'll see what happens when pockets of hot and cold air meet each other. Since it's hard to visualize air, we'll use something really cheap and easy to visualize: water!

Materials:

- Running water
- A kettle
- Some ice cubes
- 2 small pitchers or large drinking glasses
- Two different colors of food coloring (blue and yellow, blue and red, or red and yellow are the best combinations)
- Table salt
- A piece of cardboard
- Duct tape
- A rectangular clear glass or plastic container (like a baking dish or box with a lid)
- Notepaper

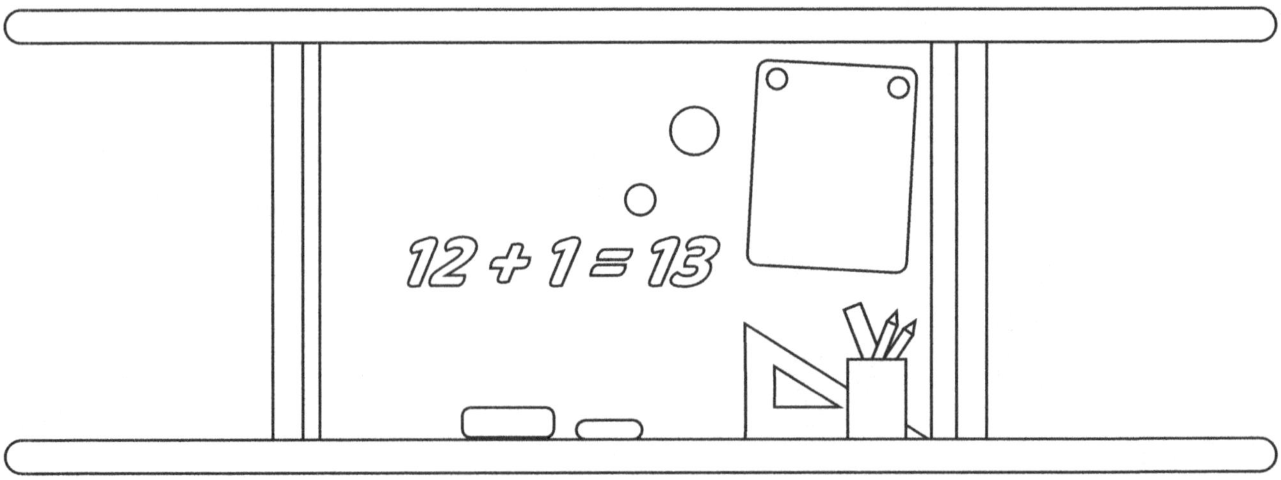

WEEK 6
DAY 6

EXPERIMENT

Procedure:

1. Measure the depth and width of your clear plastic or glass container and cut a piece of cardboard that you can use to divide the container in half, like the dividing line in the middle of a domino. Once your piece of cardboard is cut the right side, completely cover it with duct tape to waterproof it.

2. Fill one small pitcher or large drinking glass with cold tap water, a few ice cubes, and about a teaspoon of table salt. Put in a few drops of food coloring This water represents **cold air** up in the atmosphere.

3. Warm some water in the kettle (**do not use boiling water**, as that may crack your glass and create a dangerous situation) and place that hot water into the second pitcher or glass. Put in a few drops of the other food coloring. This water represents **hot air** up in the atmosphere.

4. On your notepaper, write down a few predictions about what will happen when the hot water and the cold water mix. Think about the Week 4 experiment, and make sure to brainstorm how you think the color and temperature of the water will change.

5. Place your duct tape-covered piece of cardboard into the container to divide it in half. Then, pick up one pitcher in each hand and pour them into opposite ends of the container. The hot water should be poured on one side of the divider. The cold water should be poured on the other (try not to get any ice cubes in there). As you pour, watch how the waters first begin to react to each other from across the barrier you created (it's okay if a little water is seeping underneath!).

6. Once both waters have been transferred to opposite sides of the container, slowly pull the divider out of the container and allow the two sides to interact.

7. Once the water has "mixed," observe the container from **both the top and the side** and describe what you see in your notes.

8. Answer the questions below. Then, clean up your materials.

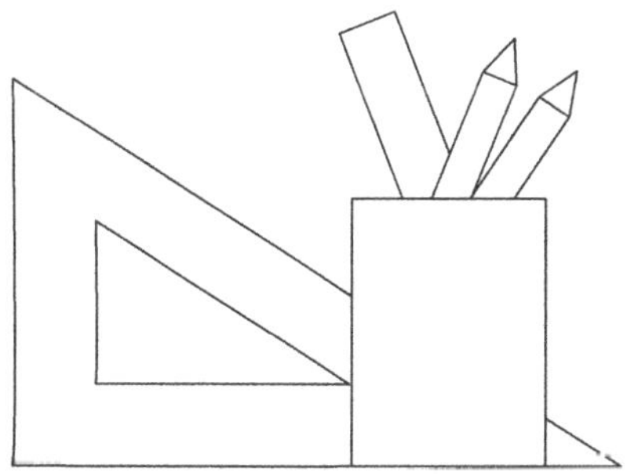

WEEK 6 DAY 6 EXPERIMENT

Follow-Up Questions:

1. Based on what you saw today, which is **heavier**: hot air or cold air? What from the experiment makes you say that?

____

____

____

2. If a lot of hot air is getting pushed upward by a lot of cold air, would people on the ground below feel hot or cold? What makes you say that?

____

____

____

YOGA

Please be aware of your environment and be safe at all times. If you cannot do an exercise, just try your best.

2 - Down Dog: 30 sec.

3 - Stretching: Stay as long as possible.
Note: do on one leg then on another.

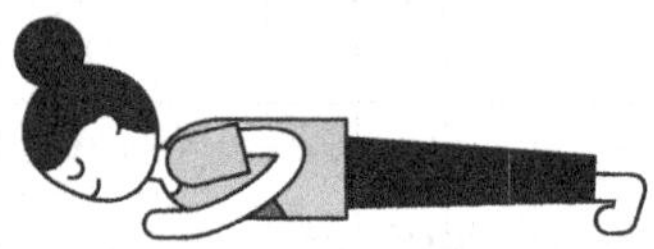

4 - Lower Plank: 20 sec.
Note: Keep your back straight and body tight.

1 - Tree Pose: Stay as long as possible.
Note: do on one leg then on another.

6 - Shavasana: 5 min.
Note: this pose is very important and provides you with long term benefits. Try not to skip this. Close your eyes and imagine who you want to be and what your goals are! Always think happy thoughts.

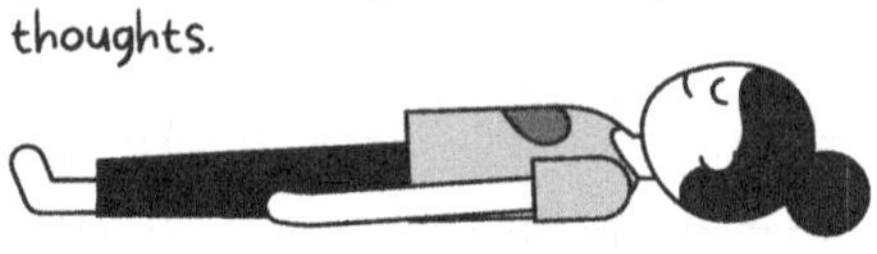

5 -Book Pose: 20 sec.
Note: Keep your core tight. Legs should be across from your eyes.

WEEK 6 DAY 7 LOGIC

Task:

All the triangular prisms are identical for each respective row.
Which row shows the back side view of row 1?

WEEK 7

ARGOPREP

OVERVIEW OF ENGLISH CONCEPTS

Reading & Writing Plays

Over the last few weeks, you've been reading Bernard Shaw's play *Pygmalion*, which was the inspiration for the famous musical *My Fair Lady*. As you've probably noticed, the fact that *Pygmalion* is a **drama** or **play** has created a very different reading experience. In some ways, reading plays is **quick** and **lively**. In other ways, it can get a little confusing. This week, we'll look at some strategies to keep your comprehension on track as you read plays.

Plays, of course, follow the same general **narrative structure** as other works of literary fiction: they begin by introducing the characters and setting, present a conflict, and then offer a series of escalating scenes that built toward a climax. With that said, plays aren't written in traditional paragraphs, sections, or chapters, so you need to use some different tools to understand them.

Key Terms

Scene: One "moment" or "episode" within a play; often a conversation or interaction between a few characters about a specific idea.

Act: One "section" or "chapter" within a play; often containing a number of scenes that happen around the same time or that share common themes.

Dialogue: The parts of the play where the characters are identified and speak.

Stage Directions: The parts of the play that describe what the set or costumes look like, how the characters move or interact, and any other visual details that can't be clearly communicated using dialogue.

Steps to Approaching a Play:

- Pay **equal attention** to both **dialogue** and **stage directions**
  - The **dialogue** obviously tells you what each character is saying
  - The **stage directions** provide information about the setting, costumes, and actions of the characters
    - If you're **seeing** the play, this isn't really a big deal for a viewer. If you're **reading** a play, however, the stage directions are one of your biggest resources!
- Keep focused on **which characters are in each scene**
  - A few side notes makes this really easy!
  - By keeping track of which characters have conversations in certain scenes, you can gain a better understanding of who knows what.
    - In **comedy** plays, this is often the source of a lot of the humor!

OVERVIEW OF ENGLISH CONCEPTS

- Don't be afraid to **read aloud or act out the scene**
    - Like **poetry**, drama often contains clever wordplay that can be tough to recognize during the first reading, especially when you're reading it silently to yourself.
    - Acting some scenes out or reading out loud helps you **hear the rhythm** of the scene, rather than just reading the words. This can make a big difference in your comprehension and understanding.
        - It's also **fun** and can break up the monotony of being deep inside a text.

- Try to **summarize each act or scene** once it is complete.
    - As we've said before, one of the best ways to understand the **whole** is the understand its **parts**.
    - If you just jot down a single sentence describing what happened in each scene, you'll create a strong study guide that will help you find information, locate specific scenes, and understand the overall flow of the play better

READING PASSAGE

From "Pygmalion"

By Bernard Shaw

LIZA [darkly] My aunt died of influenza: so they said.

MRS. EYNSFORD HILL [clicks her tongue sympathetically]!!!

LIZA [in the same tragic tone] But it's my belief they done the old woman in.

MRS. HIGGINS [puzzled] Done her in?

LIZA. Y-e-e-e-es, Lord love you! Why should she die of influenza? She come through diphtheria right enough the year before. I saw her with my own eyes. Fairly blue with it, she was. They all thought she was dead; but my father he kept ladling gin down her throat til she came to so sudden that she bit the bowl off the spoon.

MRS. EYNSFORD HILL [startled] Dear me!

LIZA [piling up the indictment] What call would a woman with that strength in her have to die of influenza? What become of her new straw hat that should have come to me? Somebody pinched it; and what I say is, them as pinched it done her in.

MRS. EYNSFORD HILL. What does doing her in mean?

HIGGINS [hastily] Oh, that's the new small talk. To do a person in means to kill them.

MRS. EYNSFORD HILL [to Eliza, horrified] You surely don't believe that your aunt was killed?

LIZA. Do I not! Them she lived with would have killed her for a hat-pin, let alone a hat.

MRS. EYNSFORD HILL. But it can't have been right for your father to pour spirits down her throat like that. It might have killed her.

LIZA. Not her. Gin was mother's milk to her. Besides, he'd poured so much down his own throat that he knew the good of it.

MRS. EYNSFORD HILL. Do you mean that he drank?

LIZA. Drank! My word! Something chronic.

MRS. EYNSFORD HILL. How dreadful for you!

WEEK 7 DAY 1

OVERVIEW OF ENGLISH CONCEPTS
READING PASSAGE

LIZA. Not a bit. It never did him no harm what I could see. But then he did not keep it up regular. [Cheerfully] On the burst, as you might say, from time to time. And always more agreeable when he had a drop in. When he was out of work, my mother used to give him fourpence and tell him to go out and not come back until he'd drunk himself cheerful and loving-like. There's lots of women has to make their husbands drunk to make them fit to live with [To Freddy, who is in convulsions of suppressed laughter] Here! what are you laughing at?

FREDDY. The new small talk. You do it so awfully well.

LIZA. If I was doing it proper, what was you laughing at? [To Higgins] Have I said anything I oughtn't?

MRS. HIGGINS [interposing] Not at all, Miss Doolittle.

LIZA. Well, that's a mercy, anyhow. [Expansively] What I always say is-

HIGGINS [rising and looking at his watch] Ahem!

LIZA [looking round at him; taking the hint; and rising] Well: I must go. [They all rise. Freddy goes to the door]. So pleased to have met you. Good-bye. [She shakes hands with Mrs. Higgins].

MRS. HIGGINS. Good-bye.

LIZA. Good-bye, Colonel Pickering.

PICKERING. Good-bye, Miss Doolittle. [They shake hands].

LIZA [nodding to the others] Good-bye, all.

1. Why is **Liza's** story about her aunt's health shocking to **Mrs. Eynsford Hill**?

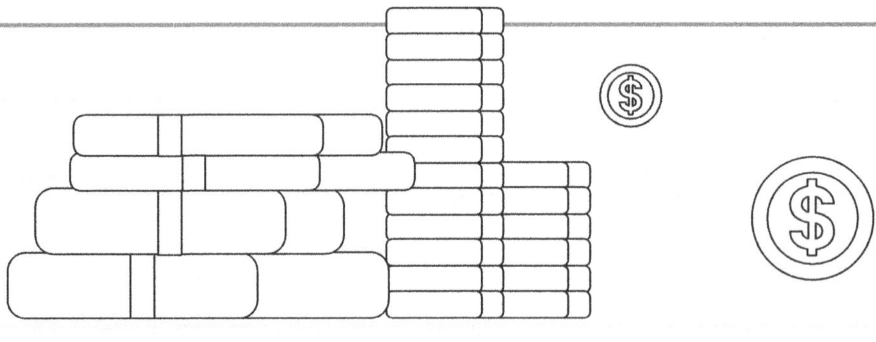

2. What is Liza's **evidence** and/or **reasoning** that leads her to believe her aunt was murdered?

3. Which of these is the closest in meaning to "the new small talk" as it is used in the passage?

   A. The hottest topics from the news
   B. The latest gossip about members of the ruling class
   C. The things people are talking about to waste time
   D. The newest slang

4. What detail about her family life does Liza reveal to the other people in the scene?

   A. Her family is filled with heavy drinkers.
   B. She has no family.
   C. All the members of her family have been murdered.
   D. She suspects her father of murdering other members of her family.

5. Do you think Freddy was **laughing at** Liza for being low class, as she suggests, or was he just **enjoying the way she explained things**, as he claims? What makes you say that? (You might want to think about previous scenes we've read as well!)

Brainstorming a Drama Scene

Directions: Today, you'll be **brainstorming** and **planning** a drama (play) scene that you will write on Day 2! We'll start with a simple **two-character** scene that involves two people talking with each other about a given conflict or issue. Use this time to think big and come up with some fun ideas that you can flesh out next time!

1. What kind of **conflict** or **conversation** will happen in your scene? (Just provide a **one-sentence summary**):

2. Who are your **two characters**?

i. Character 1's **Name**:

ii. How do they **look** or **dress**?

iii. What is their **personality** like?

iv. What's **distinct or unique** about the way they **talk** or **think**?

v. Character 2's **Name**:

vi. How do they **look** or **dress**?

OVERVIEW OF ENGLISH CONCEPTS
ACTIVITIES

vii. What is their **personality** like? ______________________________

____

____

viii. What's **distinct or unique** about the way they **talk** or **think**? ______________________________

____

____

3. What will the **setting** of this scene be like? Describe the kind of **place** it happens in. Think about how the **stage directions** in the beginning of the scenes (particularly the description of Higgins' laboratory) provide a sense of what the space looks and feels like. What will the space tell the audience about your **characters**?

____

____

____

____

FITNESS

Please be aware of your environment and be safe at all times. If you cannot do an exercise, just try your best.

Repeat these exercises **4 ROUNDS**

1 - Abs: 20 times

2 - Lunges: 8 times to each leg.
Note: Use your body weight or books as weight to do leg lunges.

3 - Plank: 20 sec.

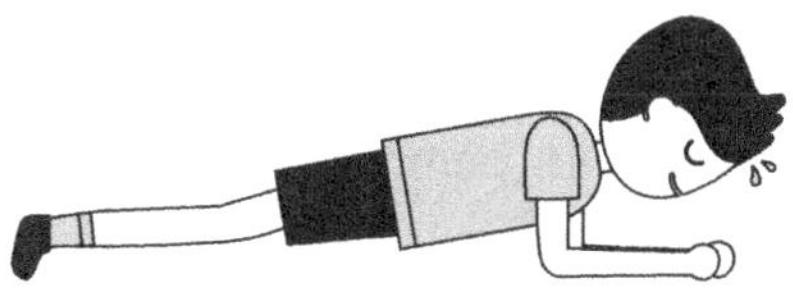

4 - Run: 50m
Note: Run **25** meters to one side and **25** meters back to the starting position.

READING PASSAGE

From "Pygmalion"

By Bernard Shaw

(Continued from Day 1's Passage, immediately after Liza leaves)

MRS. EYNSFORD HILL [suffering from shock] Well, I really can't get used to the new ways.

CLARA [throwing herself discontentedly into the Elizabethan chair]. Oh, it's all right, mamma, quite right. People will think we never go anywhere or see anybody if you are so old-fashioned.

MRS. EYNSFORD HILL. I daresay I am very old-fashioned; but I do hope you won't begin using that expression, Clara. I have got accustomed to hear you talking about men as rotters, and calling everything filthy and beastly; though I do think it horrible and unladylike. But this last is really too much. Don't you think so, Colonel Pickering?

PICKERING. Don't ask me. I've been away in India for several years; and manners have changed so much that I sometimes don't know whether I'm at a respectable dinner-table or in a ship's forecastle.

CLARA. It's all a matter of habit. There's no right or wrong in it. Nobody means anything by it. And it's so quaint, and gives such a smart emphasis to things that are not in themselves very witty. I find the new small talk delightful and quite innocent.

MRS. EYNSFORD HILL [rising] Well, after that, I think it's time for us to go.

Pickering and Higgins rise.

CLARA [rising] Oh yes: we have three at homes to go to still. Good-bye, Mrs. Higgins. Good-bye, Colonel Pickering. Good-bye, Professor Higgins.

HIGGINS [coming grimly at her from the divan, and accompanying her to the door] Good-bye. Be sure you try on that small talk at the three at-homes. Don't be nervous about it. Pitch it in strong.

CLARA [all smiles] I will. Good-bye. Such nonsense, all this early Victorian prudery!

HIGGINS [tempting her] Such nonsense!

CLARA. Such bloody nonsense!

MRS. EYNSFORD HILL [convulsively] Clara!

CLARA. Ha! ha! [She goes out radiant, conscious of being thoroughly up to date, and is heard descending the stairs in a stream of silvery laughter].

OVERVIEW OF ENGLISH CONCEPTS
READING PASSAGE

FREDDY [to the heavens at large] Well, I ask you [He gives it up, and comes to Mrs. Higgins]. Good-bye.

MRS. HIGGINS [shaking hands] Good-bye. Would you like to meet Miss Doolittle again?

FREDDY [eagerly] Yes, I should, most awfully.

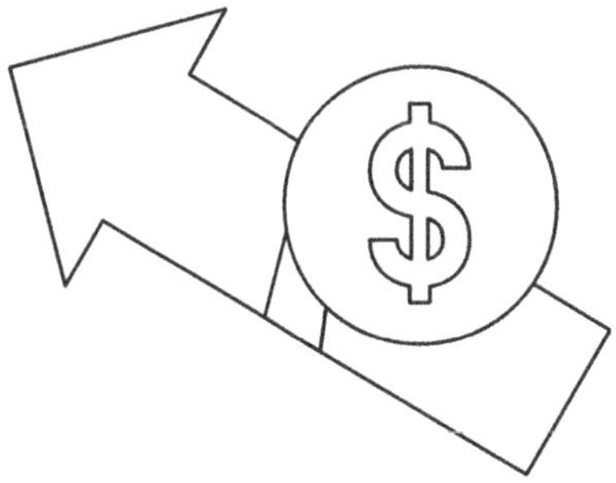

MRS. HIGGINS. Well, you know my days.

FREDDY. Yes. Thanks awfully. Good-bye. [He goes out].

MRS. EYNSFORD HILL. Good-bye, Mr. Higgins.

HIGGINS. Good-bye. Good-bye

MRS. EYNSFORD HILL [to Pickering] It's no use. I shall never be able to bring myself to use that word.

PICKERING. Don't. It's not compulsory, you know. You'll get on quite well without it.

MRS. EYNSFORD HILL. Only, Clara is so down on me if I am not positively reeking with the latest slang. Good-bye.

PICKERING. Good-bye [They shake hands].

MRS. EYNSFORD HILL [to Mrs. Higgins] You mustn't mind Clara. [Pickering, catching from her lowered tone that this is not meant for him to hear, discreetly joins Higgins at the window]. We're so poor! and she gets so few parties, poor child! She doesn't quite know. [Mrs. Higgins, seeing that her eyes are moist, takes her hand sympathetically and goes with her to the door]. But the boy is nice. Don't you think so?

1. How are **Clara** and **Mrs. Eynsford Hill's** reactions to Liza <u>different</u>?

____

____

2. Based on this scene, how do you feel about the **prediction** you made about Freddy at the end of Day 1's passage? How do you think he feels about her now?

____

____

3. According to the text, which of these is a **negative** slang term for a man?

   A. Beastly
   B. Filthy
   C. Bloody
   D. Rotter

4. What **word** does Mrs. Eynsford Hill say she'll never bring herself to use?

   A. Nonsense
   B. Bloody
   C. Good-bye
   D. Small Talk

5. How is **Clara** both similar to and different from Liza?

___

___

___

___

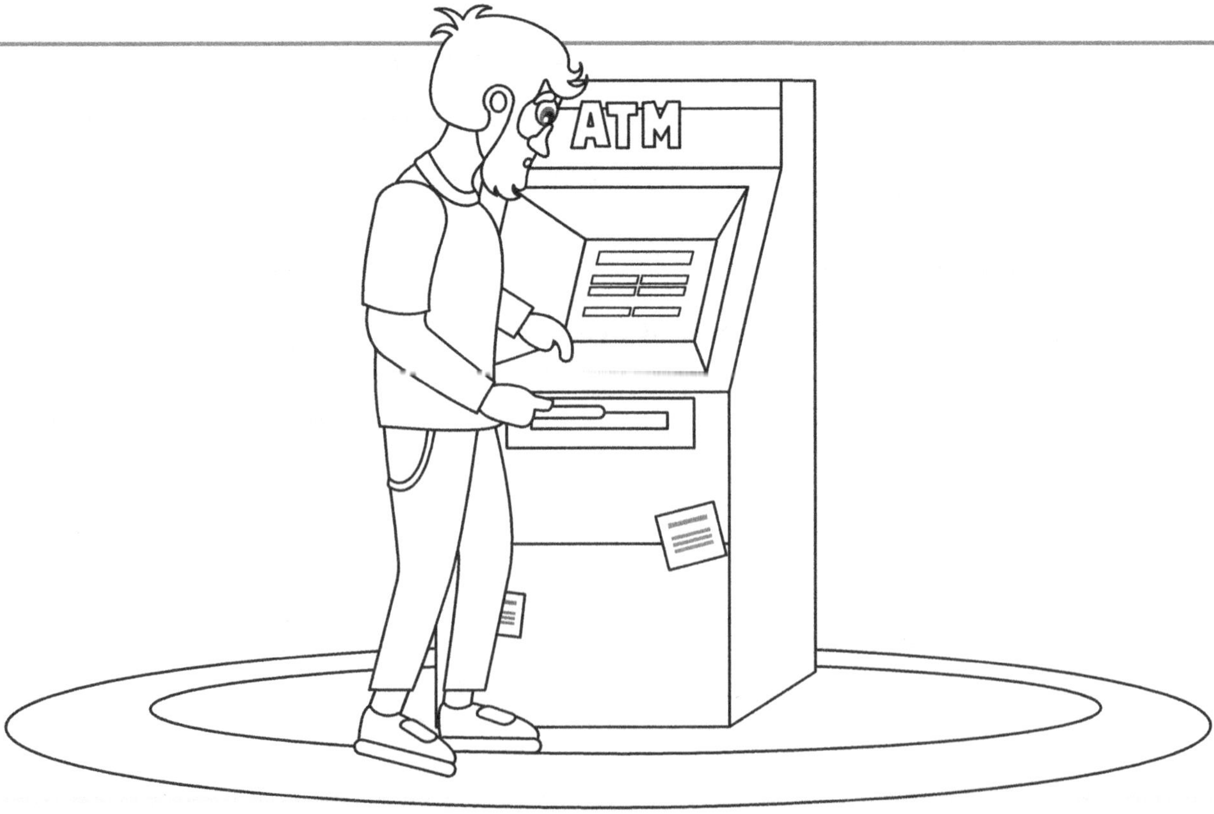

Writing a Dramatic Scene

Directions: Today, you'll be **writing** the scene that you brainstormed back on Day 1. You'll definitely want to refer to that brainstorm sheet as you work. Format your scene like we've seen done in ***Pygmalion***, with stage directions describing the scene at the beginning and individual lines of back-and-forth dialogue between the characters. Be sure your characters make it **clear** what the **conflict** or main topic of their conversation is, and try to end your scene in a memorable way!

WEEK 7
DAY 2

OVERVIEW OF ENGLISH CONCEPTS

ACTIVITIES

FITNESS

Please be aware of your environment and be safe at all times. If you cannot do an exercise, just try your best.

Repeat these exercises **4 ROUNDS**

1 - Squats: 20 times.
Note: imagine you are trying to sit on a chair.

2 - Side Bending: 15 times to each side. Note: try to touch your feet.

3 - Tree Pose:
Stay as long as possible.
Note: do the same with the other leg.

WEEK 7 DAY 3 MATH

Geometry

1. A scale on a map shows that 2 cm - 4 kilometers. If the actual distance you want to represent is 62 kilometers, how many centimeters should you draw on the map?

   A. 29 cm
   B. 30 cm
   C. 31 cm
   D. 32 cm

2. A scale of a blueprint of a park shows that 1 inch= 6 feet. If the actual distance of a ballfield that you would like to represent is 320 feet, how many inches should you draw on the map?

   A. 53 inches
   B. $53\frac{1}{3}$ inches
   C. $53\frac{2}{3}$ inches
   D. 54 inches

3. A scale on a map shows that 1 cm= 11 kilometers. If the distance on the map is 14 centimeters, what is the actual distance being represented?

   A. 154 km
   B. 154 cm
   C. 1.27 km
   D. 1.27 cm

4. A scale on a blueprint shows that 1 inch= 9 feet. If the distance on the map is 8.5 inches, what is the actual length being represented?

   A. 94 feet
   B. 94 inches
   C. 76.5 inches
   D. 76.5 feet

5. One quadrilateral has an area of 72 square feet. If you draw a new quadrilateral using a scale of 3, what is the new area of the new quadrilateral?

   A. 648 square feet
   B. 216 square feet
   C. 24 square feet
   D. 168 square feet

6. One quadrilateral has an area of 42 square inches. If one side is 6 inches, and you reproduce it using a scale factor of $\frac{1}{2}$, what is the new area?

   A. 7 square inches
   B. 10.5 square inches
   C. 21 square inches
   D. 48 square inches

7. One quadrilateral has an area of 20 square meters. If you reconstruct it using a scale factor of 4, what is the new area?

   A. 40 square meters
   B. 80 square meters
   C. 320 square meters
   D. 440 square meters

8. Use a protractor and a ruler to create a scalene triangle.

MATH

Geometry

9. Use a protractor and a ruler to create an isosceles triangle.

10. Use a protractor and a ruler to create an equilateral triangle.

11. Use a protractor and a ruler to create a right triangle.

12. Use a protractor and a ruler to create an acute triangle.

13. Use a protractor and a ruler to create an obtuse triangle.

14. Use a protractor and a ruler to create an obtuse, isosceles triangle.

FITNESS

Please be aware of your environment and be safe at all times. If you cannot do an exercise, just try your best.

Repeat these exercises **4 ROUNDS**

1 - Bend forward: 15 times.
Note: try to touch your feet. Make sure to keep your back straight and if needed you can bend your knees.

2 - Lunges: 12 times to each leg.
Note: Use your body weight or books as weight to do leg lunges.

3 - Plank: 20 sec.

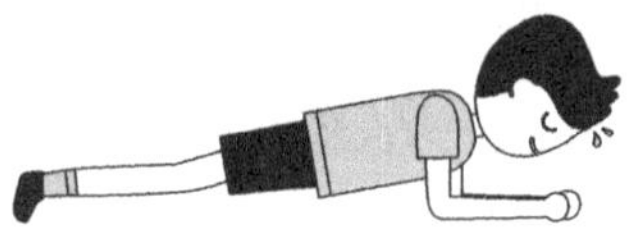

4 - Abs: 20 times

MATH

Geometry

1. What shape would be represented by the cross section below?

A. Trapezoid
B. Triangle
C. Square
D. Rectangle

2. What shape would be represented by the cross section below?

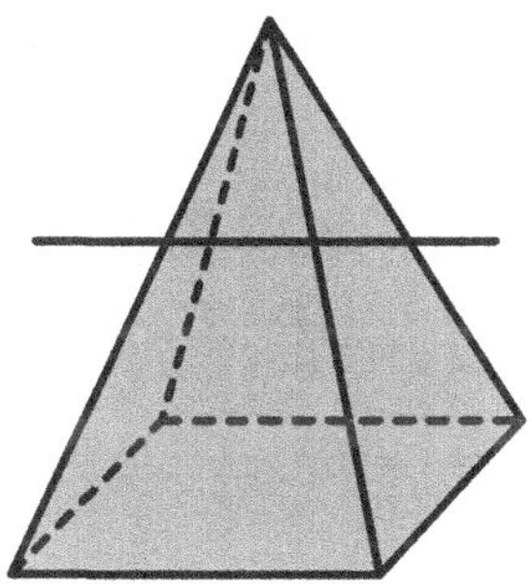

A. Trapezoid
B. Pentagon
C. Square
D. Triangle

3. What shape would be represented by the cross section below?

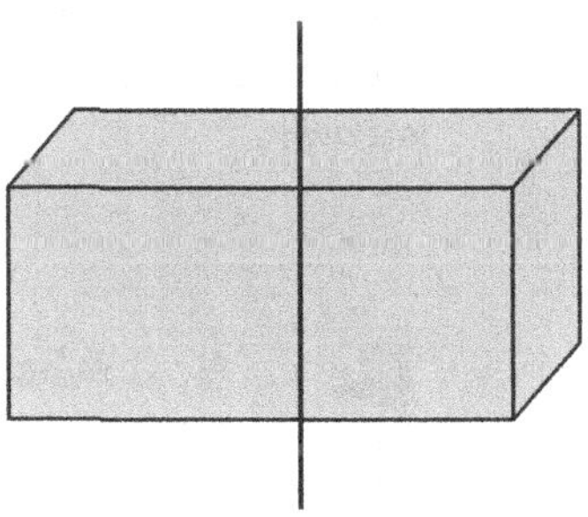

A. Triangle
B. Square
C. Pentagon
D. Circle

4. What shape would be represented by the cross section below?

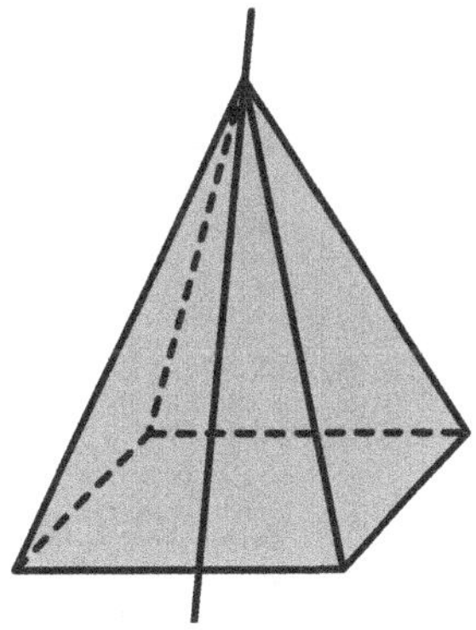

A. Triangle
B. Rectangle
C. Square
D. Pentagon

5. What shape would be represented by the cross section below?

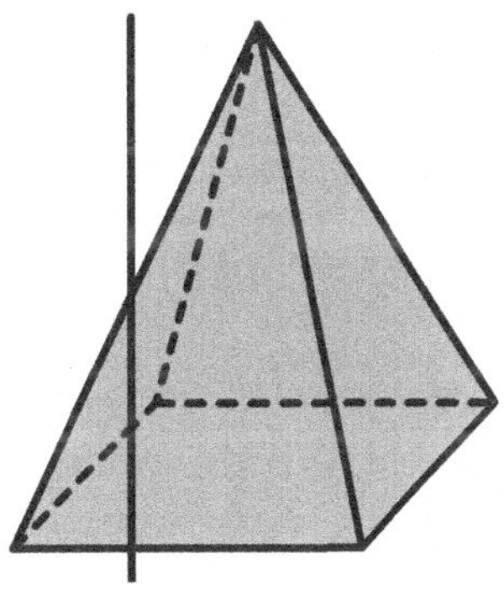

A. Trapezoid
B. Pentagon
C. Circle
D. Rectangle

6. If a circle has a diameter of 10 inches, what is its area?

A. 157.1 in^2
B. 314.1 in^2
C. 31.4 in^2
D. 78.5 in^2

7. If a circle has a radius of 6 centimeters, what is its area?

A. 18.8 cm^2
B. 37.7 cm^2
C. 113.1 cm^2
D. 452.4 cm^2

WEEK 7 DAY 4 MATH

Geometry

8. If a circle has a circumference of 22 m, what is its area?

   A. 16.2 m^2 C. 48.2 m^2
   B. 38.5 m^2 D. 72.4 m^2

9. If a circle has an area of 68 meters squared, what is its diameter?

   A. 9.3 m C. 10.9 m
   B. 10.4 m D. 11.4 m

10. If a circle has a diameter of 21 feet, what is its area?

   A. 65 ft^2 C. 346.2 ft^2
   B. 340.2 ft^2 D. 1385.4 ft^2

11. What is the radius of a circle with a circumference of 42 inches?

   A. 1.2 in C. 6.7 in
   B. 3.65 in D. 13.3 in

12. What is the circumference of a circle with a diameter of 20 meters?

   A. 62.8 m C. 314.2 m
   B. 125.7 m D. 1256.6 m

13. What is the diameter of a circle with a circumference of 89 feet?

   A. 14.2 ft C. 279.6 ft
   B. 28.3 ft D. 24885.6 ft

14. A circle has a circumference of 81.5 inches, what is its diameter?

   A. 5.1 in C. 12.9 in
   B. 8.6 in D. 25.9 in

15. A circle has an area of 124 meters squared. What is its diameter?

   A. 12.6 m C. 25.2 m
   B. 24.2 m D. 6.3 m

FITNESS

Please be aware of your environment and be safe at all times. If you cannot do an exercise, just try your best.

Repeat these exercises **4 ROUNDS**

1 - High Plank: 25 sec.

2 - Chair: 20 sec.
Note: sit on an imaginary chair, keep your back straight.

3 - Waist Hooping: 20 times. Note: if you do not have a hoop, pretend you have an imaginary hoop and rotate your hips 10 times.

4 - Abs: 25 times

WEEK 7 DAY 5 MATH

Geometry

1. What is the supplementary angle to an angle that measures 120°?

   A. 120° C. 20°
   B. 60° D. 80°

2. What is the complementary angle to an angle that measures 77°?

   A. 283° C. 103°
   B. 144° D. 13°

3. What is the measurement of a vertical angle whose opposite measures 105°?

   A. 95° C. 105°
   B. 210° D. 75°

4. The total measurement of two adjacent angles is 85°? If one angle is 9°, what is the other angles measurement?

   A. 76° C. 152°
   B. 18° D. 189°

5. Which equation would find the adjacent angle to an angle that measures 12° and makes the pair total 112°?

   A. $x - 12 = 112$
   B. $x + 12 = 112$
   C. $112 + x = 180$
   D. $90 + x = 112$

6. Which equation would find the supplementary angle to an angle that measures 31°?

   A. $x + 31 = 180$
   B. $180 - 90 = x$
   C. $x + 31 = 90$
   D. $x - 31 = 90$

7. Which equation would find the measurement of one angle of a set of vertical angles whose adjacent angles measure 64° each?

   A. $2x + 64 = 180$
   B. $x + 64 = 90$
   C. $x + 28 = 180$
   D. $2x + 128 = 360$

8. Which equation would find the complementary angle to an angle that measures 46°?

   A. $2x + 46 = 180$
   B. $90 - x = 180$
   C. $x - 46 = 180$
   D. $x + 46 - 90$

9. What is the area of a triangle with a base of $11x$ cm and a height of 6 cm?

   A. $33x$ cm^2 C. $66x$ cm^2
   B. 33 cm^2 D. 66 cm^2

10. What is the area of a trapezoid with bases of $3a$ and $14a$ and a height of 5 inches?

    A. $85a$ in^2 C. 85 in^2
    B. $42.5a$ in^2 D. 42.5 in^2

11. What is the volume of the figure below?

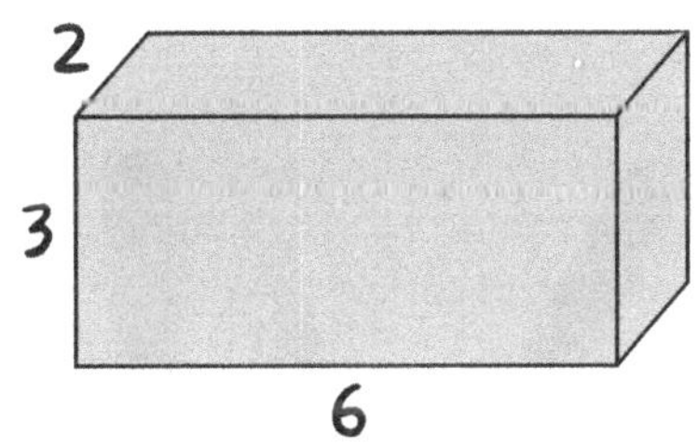

    A. 36 C. 72
    B. 16 D. 144

WEEK 7 DAY 5 MATH

Geometry

12. What is the volume of the figure below if each side measures 11 cm?

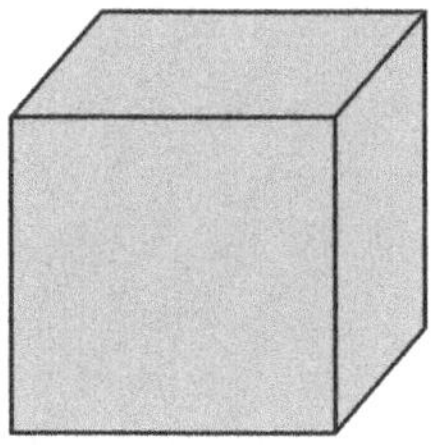

A. 33 cm^3
B. 121 cm^3
C. 1331 cm^3
D. 66 cm^3

13. If Dwight wants to wrap a birthday present for his mom and the box is 12 inches by 6 inches by 4 inches, how much area should he expect to cover?

A. 288 in^2
B. 144 in^2
C. 576 in^2
D. 432 in^2

14. What is the surface area of the figure below?

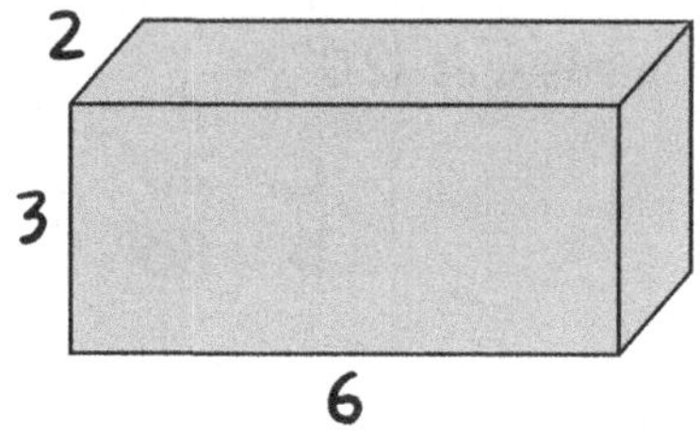

A. 36
B. 72
C. 144
D. 288

15. What is the surface area of the figure below if each side measures 11 cm?

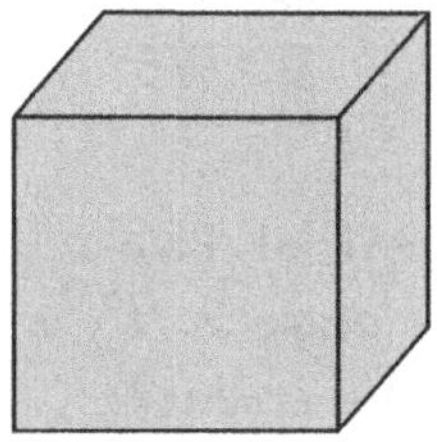

A. 726 cm^2
B. 121 cm^2
C. 1331 cm^2
D. 847 cm^2

YOGA

Please be aware of your environment and be safe at all times. If you cannot do an exercise, just try your best.

1 - Down Dog: 30 sec.

2 - Bend Down: 30 sec.

3 - Chair: 30 sec.

4 - Child Pose: 30 sec.

5 - Shavasana: as long as you can. Note: think of happy moments and relax your mind.

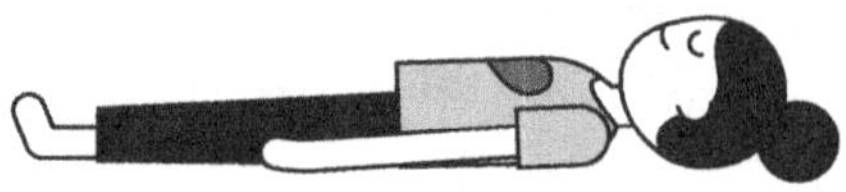

WEEK 7 DAY 6 EXPERIMENT

Tornadoes

*Last week, you saw how pockets of hot and cold air contribute to the temperature and weather by partially mixing and pushing each other around. One of the most extreme weather events that can occur when conditions are just right is a **tornado**. A **tornado** is a spinning vortex of winds that can form as part of a large, powerful weather system.*

*Today, you'll use **water** to simulate **air**, just like we did in the Week 6 experiment. This time, however, you'll see how air can stir up to form a vortex and cause a **tornado**!*

Materials:

- 2 empty, clean two-liter bottles
- Running water
- 1 two-inch long piece of half-inch PVC pipe (available at the hardware store)
- A hot glue gun (with glue sticks)
- An adult
- Food coloring (optional)

Procedure:

1. Double check that your two-liter bottles are clean. If the labels are still attached, remove them so you can see right through both bottles.

2. Fill one bottle about 2/3 full with water. Try to visualize where the top of the label used to be, if you need a reference mark. If you have some food coloring, add a few drops and stir them in to make the demonstration more dramatic.

3. Plug in your hot glue gun and allow it to warm up. **If you've never used a hot glue gun before, you should get an adult at this point. Hot glue (and the metal parts of the gun) can give you a nasty burn if you're not careful!**

4. Once your glue gun has warmed up, tilt the bottle with the water in it slightly and put a ring of hot glue down into the neck of the bottle, near where the bottom edge of the cap would be. **Have an adult do this if you're not experienced using the hot glue gun.**

5. As soon as your glue ring is in the neck of the bottle, take your two-inch long piece of PVC pipe and slide it down into the neck of the bottle (this will probably involve some twisting and firm pressing), so that **half** of the pipe is in the bottle and **half** is poking above the top.

WEEK 7 DAY 6 EXPERIMENT

6. Put a second ring of hot glue around the top edge of the second empty bottle (where you would put your mouth), and quickly press it down onto the top of the pipe, holding things firmly together for a minute or so to allow the glue to harden and a seal to form. Once this step is completed, you should have an assembly that looks like two bottles attached at their mouths with a piece of pipe reinforcing that connection. **Have an adult do this if you're not experienced using the hot glue gun.**

7. Once your glue is dry and your seal is firm, grasp the two bottles by the middle and hold them so that the bottle with the liquid in it is on top and upside down. Notice the weak flow of both **air** and **water** between the two bottles.

8. With the filled bottle on top, spin the bottle assembly around like you were stirring a big pot, creating a spinning vortex in the chamber. Observe the vortex as it forms and travels to the bottom bottle.

9. Flip your two-bottle assembly back over and repeat this process a few times, varying the speed and force of your spinning. Try to see if you can control the shape and size of the vortex by applying slightly different strategies.

10. Once you've mastered creating different vortexes, clean up your materials **(be especially careful handling the glue gun)**, and answer the questions below.

WEEK 7
DAY 6

EXPERIMENT

Follow-Up Questions:

1. How was the **water** in the bottle acting like the **air or wind** in a tornado?

2. Based on what you saw today, why do you think it sometimes **rains very hard** during a tornado?

YOGA

Please be aware of your environment and be safe at all times. If you cannot do an exercise, just try your best.

3 - Stretching: Stay as long as possible.
Note: do on one leg then on another.

2 - Down Dog: 30 sec.

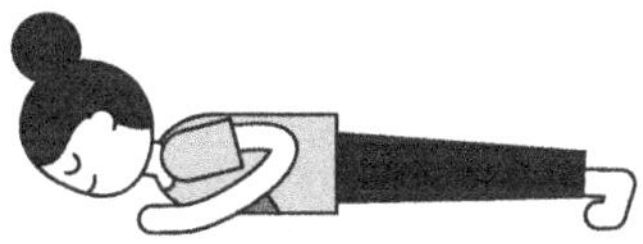

4 - Lower Plank: 20 sec.
Note: Keep your back straight and body tight.

1 - Tree Pose: Stay as long as possible.
Note: do on one leg then on another.

6 - Shavasana: 5 min.
Note: this pose is very important and provides you with long term benefits. Try not to skip this. Close your eyes and imagine who you want to be and what your goals are! Always think happy thoughts.

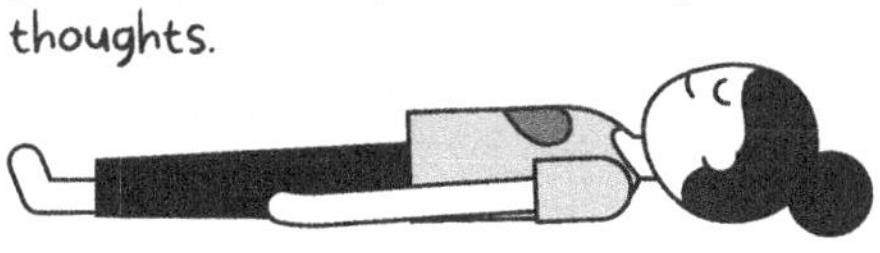

5 -Book Pose: 20 sec.
Note: Keep your core tight. Legs should be across from your eyes.

WEEK 7 DAY 7 LOGIC

Task:

Which of the following images are cubes? Hint: Look closely as some images look like cubes but they are not.

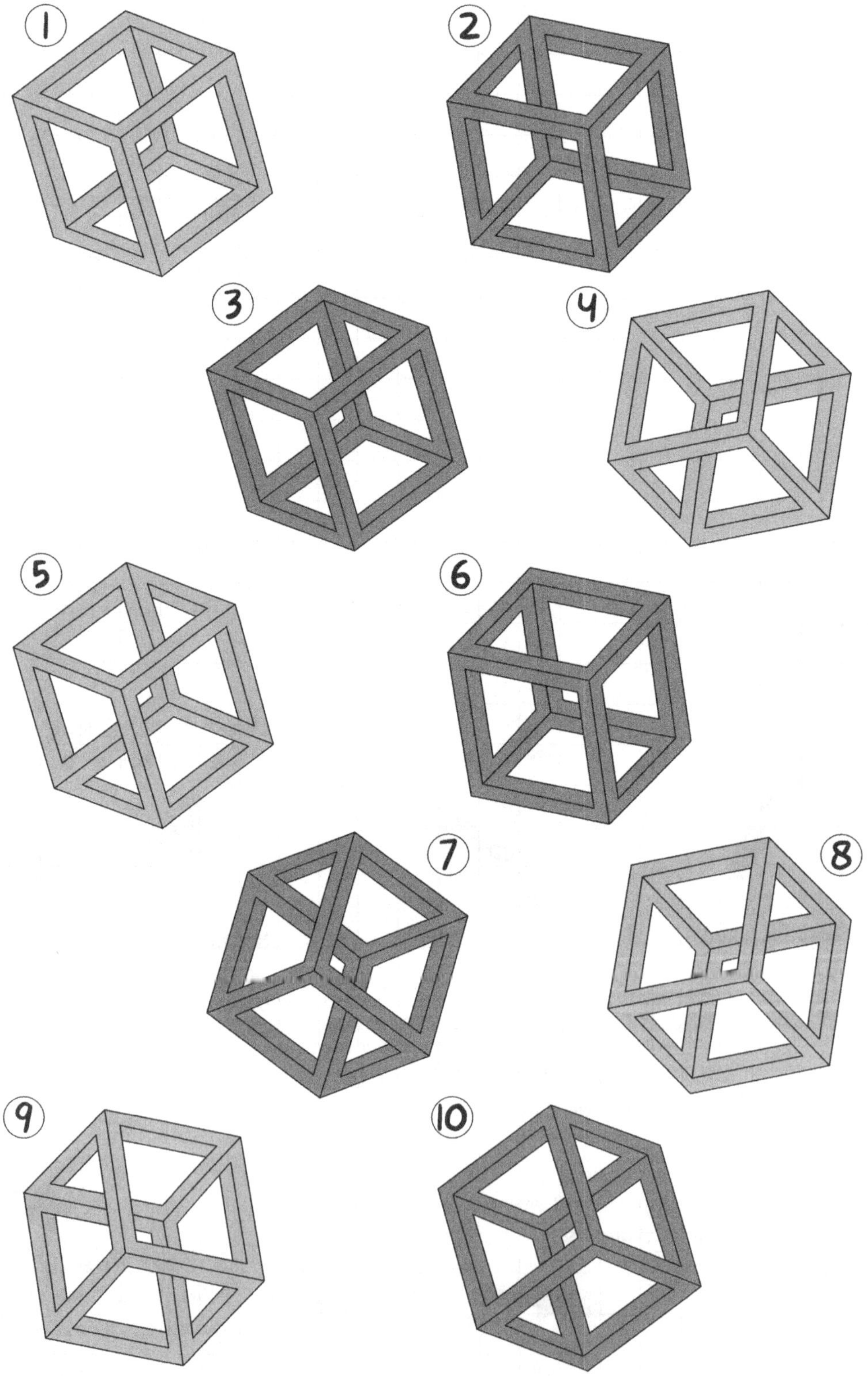

WEEK 8

ARGOPREP

WEEK 8 DAY 1 OVERVIEW OF ENGLISH CONCEPTS

Observing Structure & Organization in Informational Texts

Over the last few weeks, we've been focusing on examples of **literature** in both the myths of ancient Greece and the play *Pygmalion*. Now, it's time to shift gears and think about **informational texts**. Remember, the goal of an informational text is to **communicate an idea** or **discuss a topic** in a way that provides the reader with as much valuable, clear, and relevant information as possible.

One of the main ways authors provide a **clear**, easy-to-read experience is by relying on **structure** and **organization**. If you feel lost in an informational text, try thinking about the **structure** of the text rather than the content: What is the name of the chapter? Are there any nearby section heads? What ideas has the author already talked about? While these questions may seem only tangentially related to reading comprehension, they're actually quite valuable when it comes to informational texts!

Key Terms:

Introduction: The part at the beginning of a text that lays out main ideas and helps the reader understand what they're reading and why it's relevant or important.

Body: The part in the middle of a text, where the author presents individual points or ideas one at a time, providing examples to clarify and working to tie ideas together.

Conclusion: The part at the end of a text, where the author recaps main ideas, reminds the reader of the relevance of what they read, and wraps things up.

Strategies for Reading with Structure in Mind:

1. When reading the beginning of an informational text, try to separate the **introduction** from the rest of the text.
    a. In a **short text**, the introduction might only be a few sentences or a paragraph or two. In a **long text**, the introduction might be an entire section or chapter.
    b. Once you've found the beginning and the end of the **introduction**, read them through again **on their own** – don't even think about the rest of the text yet.
        i. Ask yourself: **What ideas** is the author presenting? What should I **expect to hear**? What **predictions can I make** based on what I've already read?
    c. Once you've built a strong understanding of the **introduction**, you can approach the body in a much more informed way.

2. When you're reading the **body** of a text, try to identify each new **point or idea** that the author introduces and ask yourself...
   a. How does this connect to the information I read in the **intro**?
   b. How is this point connected to the one **that came before it**?
   c. How does this point **change or complicate my understanding** of what's happened so far?
   d. How does the author group related ideas into **paragraphs, subheadings, or chapters**?

3. When you're reading the **conclusion** of a text, keep your notes nearby and...
   a. Make sure that nothing in the conclusion seems **new** to you
      i. If it does, you probably missed something in the body and should **go back!**
   b. Ensure you have **full understanding** of all the main points the author restates in the conclusion
   c. Think about whether the author's conclusion really matches what they said in the intro and the body
      i. If not, it might be a weak conclusion, or the author may be trying to be deceptive!

As a Writer...

When you write informational texts, structure is your friend! Remember to...

- Use clear **titles** and **headings** for different sections to help orient the reader and ensure they know what's going on.
- Write **focused paragraphs** that are all about **one idea** - If something is a new idea or not connected to the other ideas in the paragraph, you should pull it out and start a new paragraph!
- Be sure to use your **introduction and conclusion** to create a "roadmap" or "guidebook" for the reader that provides as much clarity as possible in terms of what's going on and how you're presenting it.

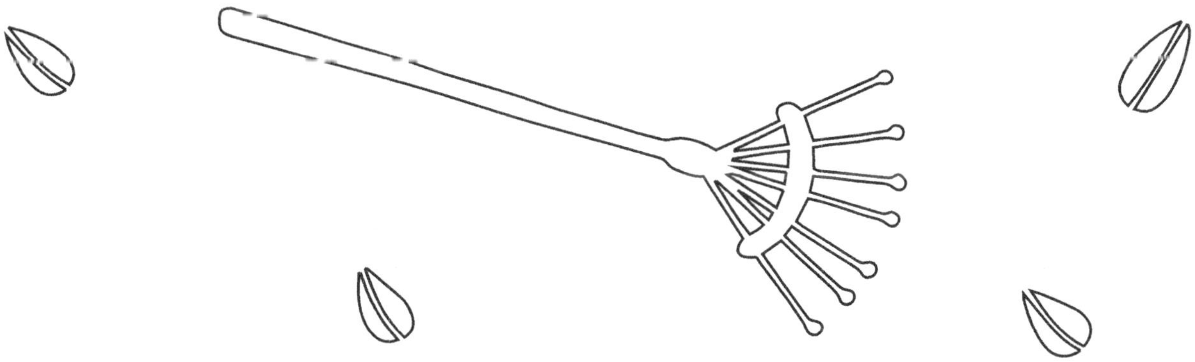

READING PASSAGE

From "The Narrative of the Life of Frederick Douglass, An American Slave"

By Frederick Douglass

I was born in Tuckahoe, near Hillsborough, and about twelve miles from Easton, in Talbot county, Maryland. I have no accurate knowledge of my age, never having seen any authentic record containing it. By far the larger part of the slaves know as little of their ages as horses know of theirs, and it is the wish of most masters within my knowledge to keep their slaves thus ignorant. I do not remember to have ever met a slave who could tell of his birthday. They seldom come nearer to it than planting-time, harvest-time, cherry-time, spring-time, or fall-time. A want of information concerning my own was a source of unhappiness to me even during childhood. The white children could tell their ages. I could not tell why I ought to be deprived of the same privilege. I was not allowed to make any inquiries of my master concerning it. He deemed all such inquiries on the part of a slave improper and impertinent, and evidence of a restless spirit. The nearest estimate I can give makes me now between twenty-seven and twenty-eight years of age. I come to this, from hearing my master say, some time during 1835, I was about seventeen years old.

My mother was named Harriet Bailey. She was the daughter of Isaac and Betsey Bailey, both colored, and quite dark. My mother was of a darker complexion than either my grandmother or grandfather.

My father was a white man. He was admitted to be such by all I ever heard speak of my parentage. The opinion was also whispered that my master was my father; but of the correctness of this opinion, I know nothing; the means of knowing was withheld from me. My mother and I were separated when I was but an infant — before I knew her as my mother. It is a common custom, in the part of Maryland from which I ran away, to part children from their mothers at a very early age. Frequently, before the child has reached its twelfth month, its mother is taken from it, and hired out on some farm a considerable distance off, and the child is placed under the care of an old woman, too old for field labor. For what this separation is done, I do not know, unless it be to hinder the development of the child's affection toward its mother, and to blunt and destroy the natural affection of the mother for the child. This is the inevitable result.

1. Based on the **limited** information included in the passage, **about** what year was Frederick Douglass born in?

2. Why is Douglass' **birthday** a source of emotional distress?

3. Which of these best represents Frederick Douglass' understanding of who his parents were?

A. He knew for certain who both his parents were.
B. He didn't know for certain who either of his parents were.
C. He knew who his mother was for sure and drew some reasonable conclusions as to who his father was.
D. He knew who his father was for sure and drew some reasonable conclusions about who his mother was.

4. Why does Frederick believe slave mothers were separated from their babies?

A. To sell the babies to other slave holders faster
B. To limit the strength of emotional bonds between slaves
C. To psychologically torture the mother
D. To prevent the baby from knowing who its real mother is

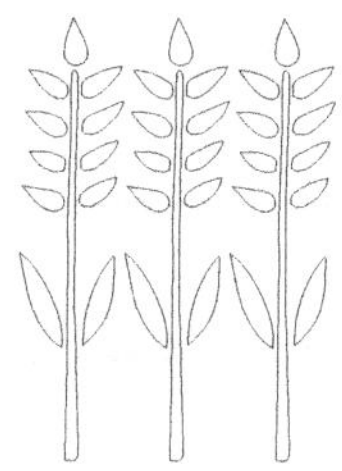

5. Based on the text, how can the reader tell that **conditions improve** for Frederick, even though his early childhood was horrific?

Reading with an Eye Toward Organization

Directions: Find a story about a topic that's interesting to you in a news-related text, such as a newspaper, magazine, or online news site. Once you've chosen an article, read it **slowly** and **carefully** with an eye toward both **what it says** and **how the author groups ideas and presents things in a certain order.** Fill out the sheet below to analyze the article in terms of structure and meaning!

(**NOTE:** There is space below to summarize up to **5** paragraphs. If the article is longer than that, you can create further summaries on another sheet of paper).

Headline:

- What's the main topic of this article? ______________________________

- PARAGRAPH 1 SUMMARY: ______________________________

- PARAGRAPH 2 SUMMARY: ______________________________

- PARAGRAPH 3 SUMMARY: ______________________________

- PARAGRAPH 4 SUMMARY: ______________________________

- PARAGRAPH 5 SUMMARY: ______________________________

OVERVIEW OF ENGLISH CONCEPTS

ACTIVITIES

1. Which paragraph helped you understand **the main idea** the best?

   ____

   - What about that paragraph was **most helpful**? ______________________

     ____

2. Did any paragraphs **stray or wander** from the main idea that was presented?

   - Which ones? ____
   - **How** did they stray or wander from the focus? ______________________

     ____

     ____

OVERVIEW OF ENGLISH CONCEPTS ACTIVITIES

3. How would you describe the way the text was **organized**? How did the author **present things** or **group ideas together**? ____

____

____

4. Write an **overall summary** of the article you read: ______________________

____

____

____

FITNESS

Please be aware of your environment and be safe at all times. If you cannot do an exercise, just try your best.

Repeat these exercises **4 ROUNDS**

2 - Lunges: 8 times to each leg.
Note: Use your body weight or books as weight to do leg lunges.

1 - Abs: 20 times

3 - Plank: 20 sec.

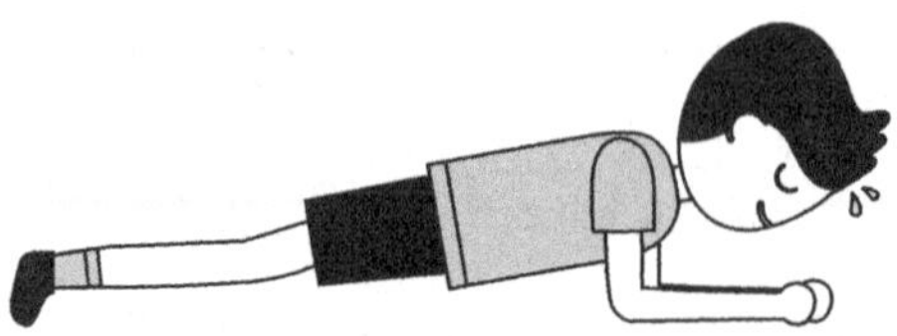

4 - Run: 50m
Note: Run 25 meters to one side and 25 meters back to the starting position.

READING PASSAGE

From "The Narrative of the Life of Frederick Douglass, An American Slave"

By Frederick Douglass

(Continued from Day 1's Passage)

I never saw my mother, to know her as such, more than four or five times in my life; and each of these times was very short in duration, and at night. She was hired by a Mr. Stewart, who lived about twelve miles from my home. She made her journeys to see me in the night, travelling the whole distance on foot, after the performance of her day's work. She was a field hand, and a whipping is the penalty of not being in the field at sunrise, unless a slave has special permission from his or her master to the contrary — a permission which they seldom get, and one that gives to him that gives it the proud name of being a kind master. I do not recollect of ever seeing my mother by the light of day. She was with me in the night. She would lie down with me, and get me to sleep, but long before I waked she was gone. Very little communication ever took place between us. Death soon ended what little we could have while she lived, and with it her hardships and suffering. She died when I was about seven on one of my master's farms, near Lee's Mill. I was not allowed to be present during her illness, at her death, or burial. She was gone long before I knew anything about it. Never having enjoyed, to any considerable extent, her soothing presence, her tender and watchful care, I received the tidings of her death with much the same emotions I should have probably felt at the death of a stranger.

Called thus suddenly away, she left me without the slightest intimation of who my father was. The whisper that my master was my father, may or may not be true; and, true or false, it is of but little consequence to my purpose whilst the fact remains, in all its glaring odiousness, that slaveholders have ordained, and by law established, that the children of slave women shall in all cases follow the condition of their mothers; and this is done too obviously to administer to their own lusts, and make a gratification of their wicked desires profitable as well as pleasurable; for by this cunning arrangement, the slaveholder, in cases not a few, sustains to his slaves the double relation of master and father.

I know of such cases; and it is worthy of remark that such slaves invariably suffer greater hardships, and have more to contend with, than others. They are, in the first place, a constant offence to their mistress. She is ever disposed to find fault with them; they can seldom do any thing to please her; she is never better pleased than when she sees them under the lash, especially when she suspects her husband of showing to his mixed-race children favors which he withholds from his black slaves. The master is frequently compelled to sell this class of his slaves, out of deference to the feelings of his white wife; and, cruel as the deed may strike any one to be, for a man to sell his own children to human flesh-mongers, it is often the dictate of humanity for him to do so; for, unless he does this, he must not only whip them himself, but must stand by and see one white son tie up his brother, of but few shades darker complexion than himself, and ply the gory lash to his naked back; and if he lisp one word of disapproval, it is set down to his parental partiality, and only makes a bad matter worse, both for himself and the slave whom he would protect and defend.

READING PASSAGE

1. Why did Frederick Douglass only see his mother **at night**?

2. Why did Douglass react to his mother's death like "**the death of a stranger**"?

3. Why does Douglass believe that, practically speaking, it didn't matter who his father was?

   A. Because his father obviously didn't care to identify himself, so he might has well have been anybody (or nobody at all)
   B. Because being born to a slave mother makes someone a slave no matter what
   C. His family probably would've just been split up by the masters anyway.
   D. He didn't have any interest in family.

4. According to Douglass, how are many masters "double relations" to their slaves?

   A. They are both owner and father.
   B. They are both owner and caretaker.
   C. They are both owner and boss.
   D. They are both owner and tormenter.

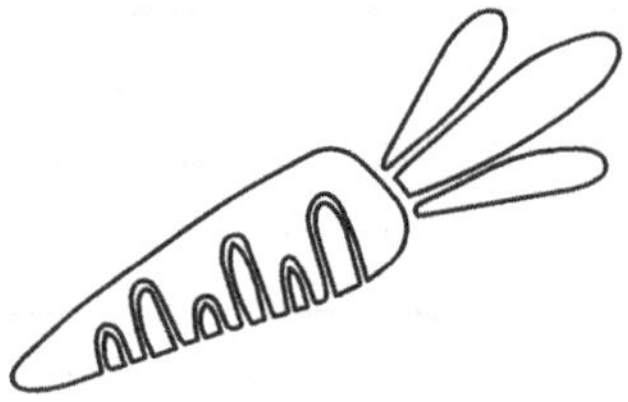

5. According to Douglass, how does a master having **slave children** create a bad situation for **many different people**?

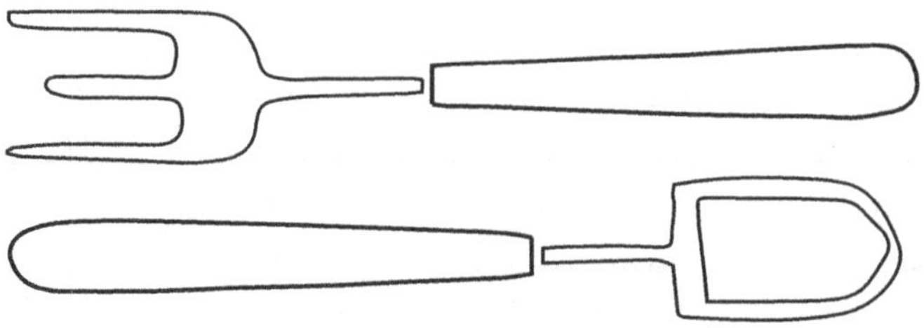

Practicing Structure & Organization

Directions: Today, you'll be **writing** a short response essay about ***The Narrative of Frederick Douglass.*** Your job isn't just to think about the text, though. You need to focus on **organization as well**. Here's how you'll do that:

Begin your essay by clearly stating your position or perspective on the prompt. Then write **two organized, focused paragraphs** explaining your thinking and reasoning. Make sure that each paragraph has a clear, specific focus. After you've finished writing a paragraph, use space in the margin to **craft a short summary** that explains the **focus** of the paragraph.

PROMPT: In what ways has Douglass shown the **dehumanizing** effects of slavery in the passages so far?

FITNESS

Please be aware of your environment and be safe at all times. If you cannot do an exercise, just try your best.

Repeat these exercises **4 ROUNDS**

1 - Squats: 20 times.
Note: imagine you are trying to sit on a chair.

2 - Side Bending: 15 times to each side. Note: try to touch your feet.

3 - Tree Pose: Stay as long as possible.
Note: do the same with the other leg.

WEEK 8 DAY 3 MATH

Geometry

1. Find the area of the figure below.

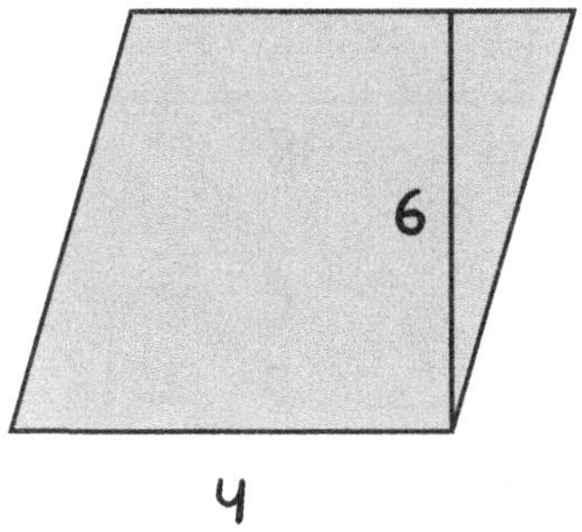

A. 24 **B.** 28 **C.** 42 **D.** 38

2. Find the area of the isosceles trapezoid below.

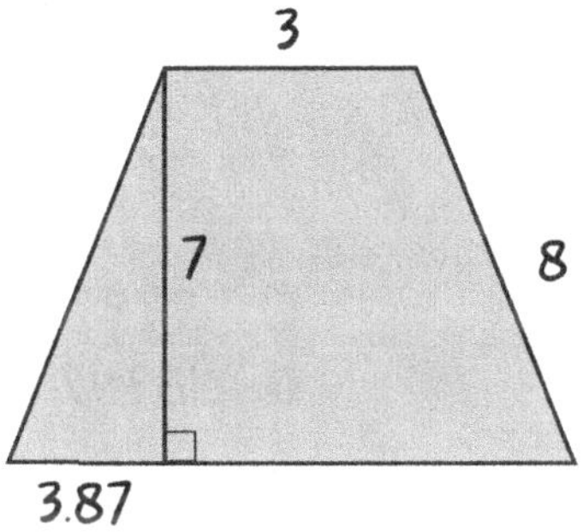

A. 43 **B.** 41.5 **C.** 40 **D.** 48.1

3. Find the area of the figure below.

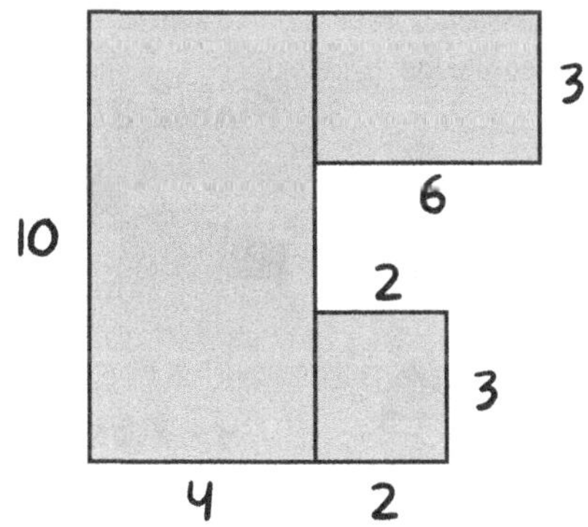

A. 26
B. 48
C. 64
D. 82

4. Find the area of the figure below.

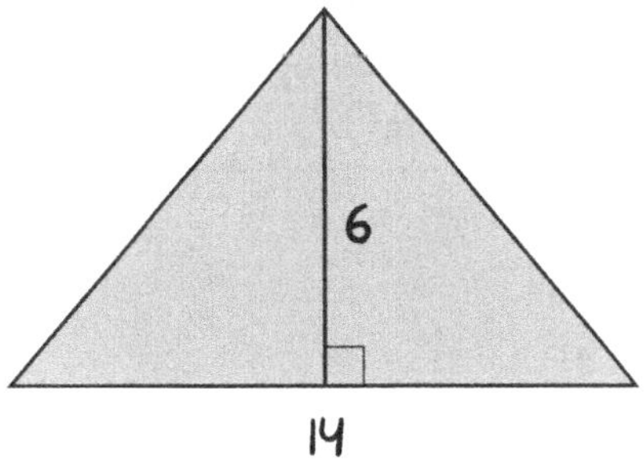

A. 36 **B.** 38 **C.** 40 **D.** 42

5. Find the area of the figure below using 3.14 as the value for π.

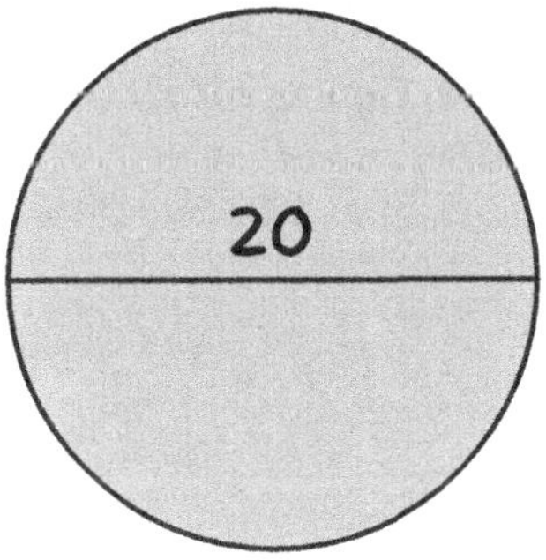

A. 314
B. 628.4
C. 31.16
D. 62.84

6. What is the missing value of a circle with a diameter of 24 cm?

A. r = 12 cm
B. r = 48 cm
C. C = 150.72 cm
D. C = 301.44 cm

7. What is the missing value of a circle with a radius of 26.5 m?

A. d = 79.2 m
B. d = 13.25 m
C. A = 2205.1 m^2
D. A = 166.42 m^2

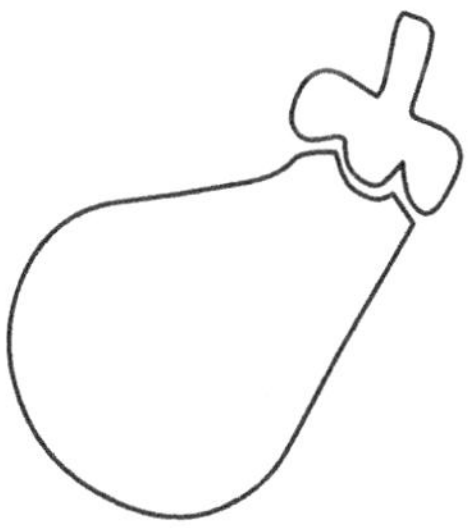

WEEK 8 DAY 3 MATH

Geometry

8. What is the missing value of a circle with an area of 8490.56 in^2?

A. r = 26 in
B. r = 208 in
C. D = 104 in
D. D = 52 in

9. What is the missing value of a circle with a circumference of 8π cm?

A. r = 16 cm
B. r = 8 cm
C. D = 16 cm
D. D = 8 cm

10. Find the area of the figure below.

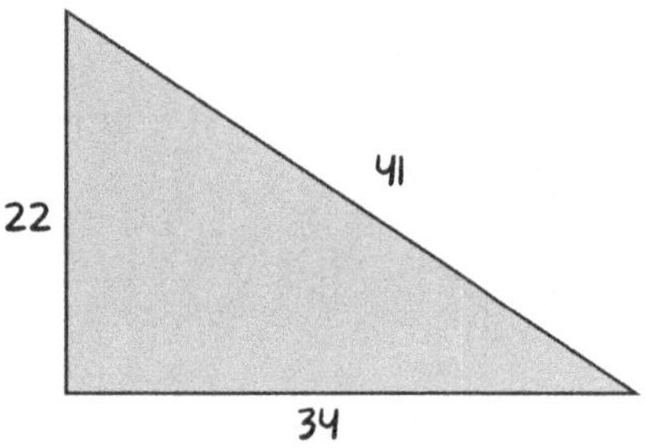

A. 37.4
B. 437
C. 374
D. 734

11. Find the area of the figure below.

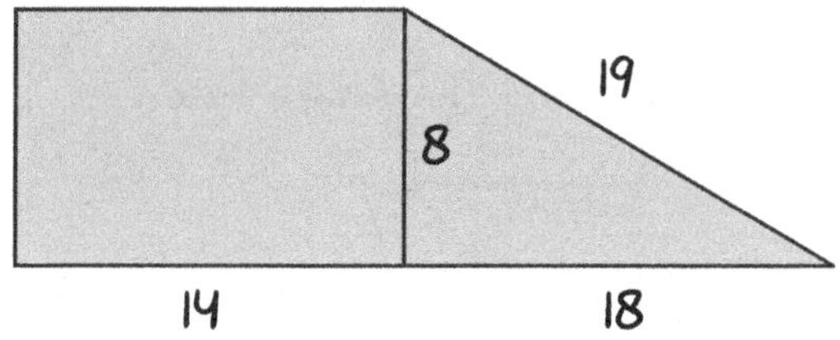

A. 184
B. 202
C. 18.4
D. 28.1

12. Find the area of the figure below using 3.14 as the value for π.

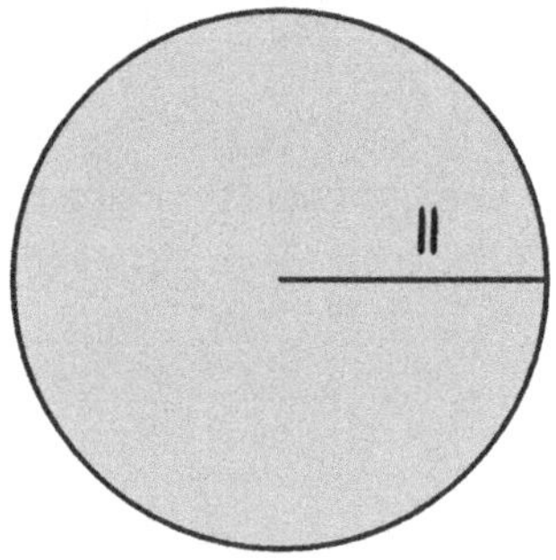

A. 1519.76
B. 379.94
C. 94.99
D. 189.97

FITNESS

Please be aware of your environment and be safe at all times. If you cannot do an exercise, just try your best.

Repeat these exercises
4 ROUNDS

1 - Bend forward: 15 times.
Note: try to touch your feet. Make sure to keep your back straight and if needed you can bend your knees.

2 - Lunges: 12 times to each leg.
Note: Use your body weight or books as weight to do leg lunges.

3 - Plank: 20 sec.

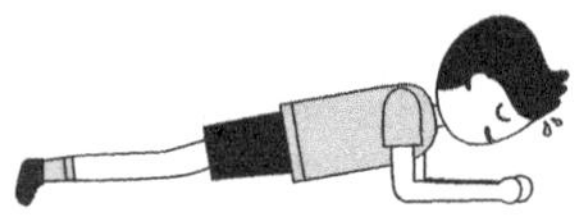

4 - Abs: 20 times

WEEK 8 DAY 4 MATH

Geometry

1. Find the surface area of the right circular cone below using 3.14 as the value for π. Hint: You can use the formula $\pi r^2 + \pi rl$ to calculate the surface area, where r is the radius and l is the slant length.

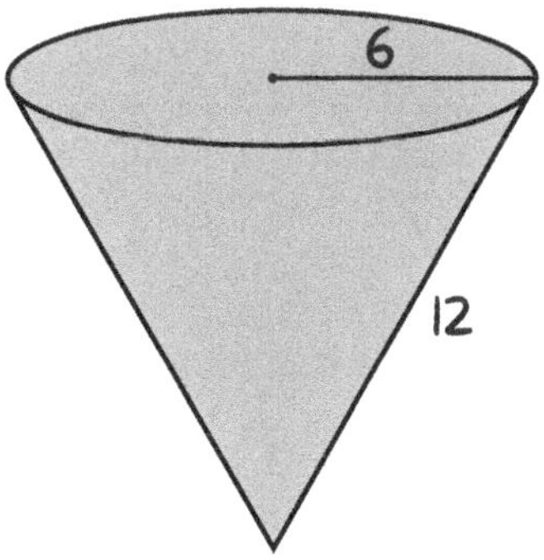

A. 348.84 **C.** 391.72
B. 339.12 **D.** 422.81

2. Find the surface area of the figure below.

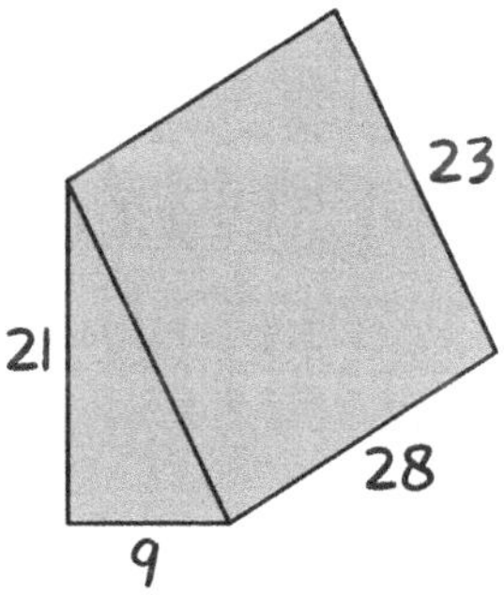

A. 967.99 **C.** 1673
B. 667.22 **D.** 1762.99

3. Find the surface area of the cylinder below using 3.14 as the value for π.

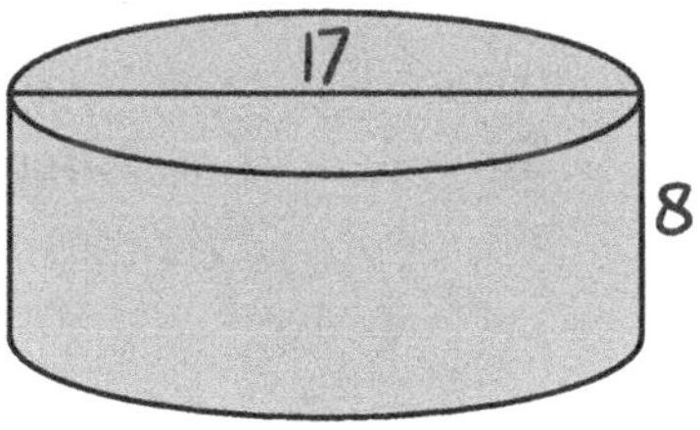

A. 840.16 **C.** 910.17
B. 880.77 **D.** 963.27

4. Find the surface area of the figure below.

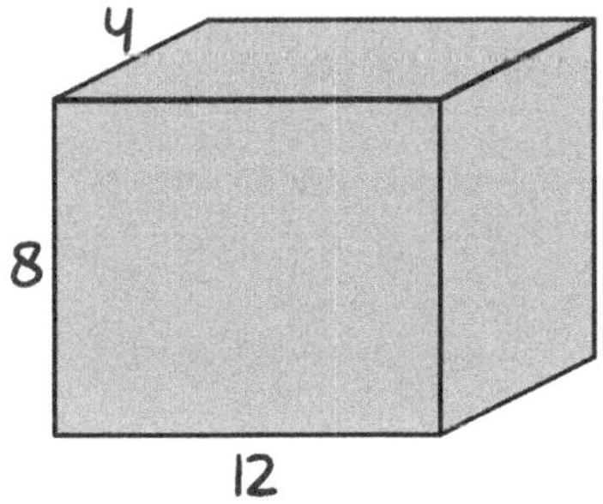

A. 168 **C.** 352
B. 176 **D.** 1,872

5. Find the volume of the figure below.

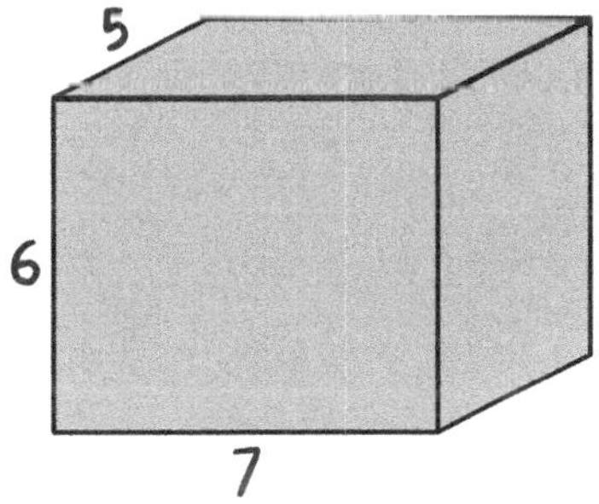

A. 190 **C.** 230
B. 210 **D.** 250

6. Find the volume of the figure below using 3.14 as the value for π.

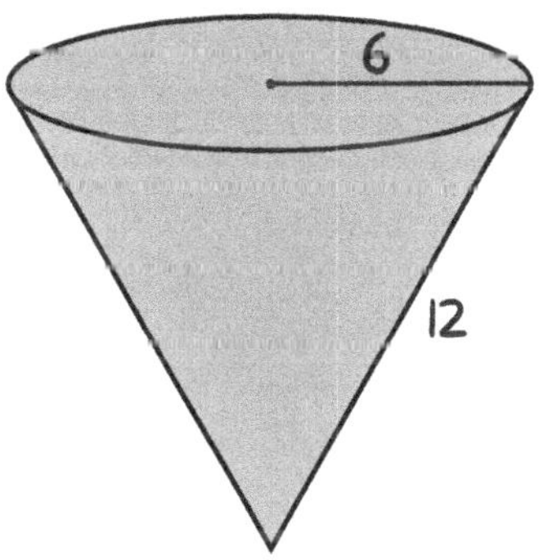

A. 391.50 **C.** 316.60
B. 316.96 **D.** 361.96

MATH

Geometry

7. Find the hypotenuse labeled c.

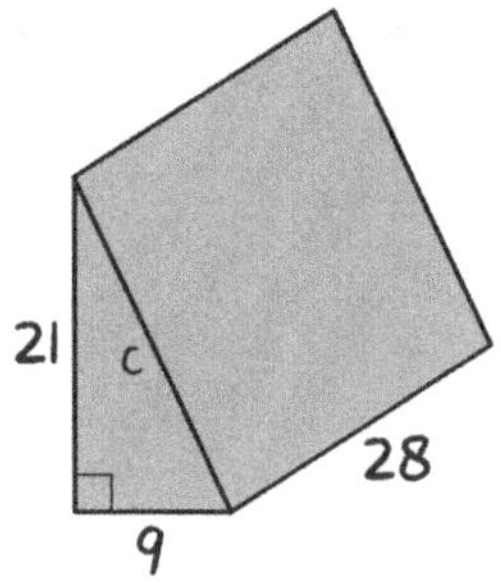

A. 20
B. 30
C. 23.2
D. 22.85

8. Find the volume of the cylinder below using 3.14 as the value for π.

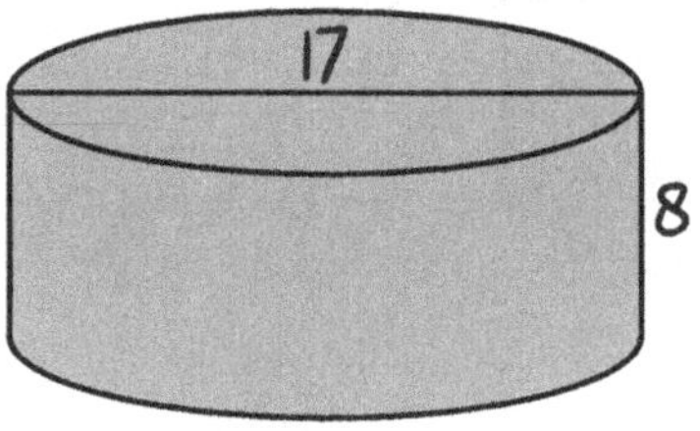

A. 815.1
B. 1518.8
C. 1814.92
D. 5881.1

9. Find the volume of the sphere below using 3.14 as the value for π.

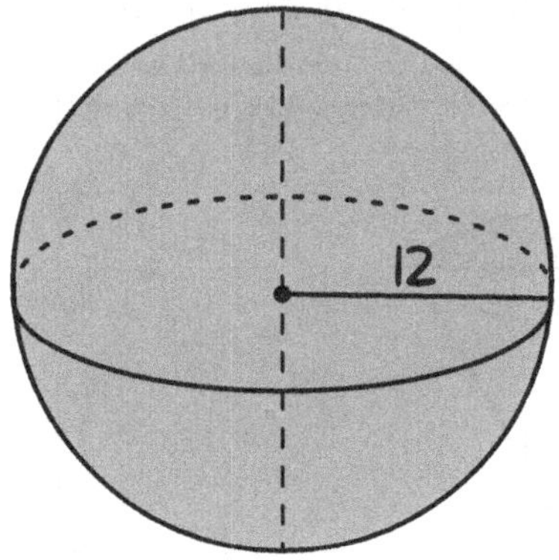

A. 7234.56
B. 7134.56
C. 7244.56
D. 6234.56

FITNESS

Please be aware of your environment and be safe at all times. If you cannot do an exercise, just try your best.

Repeat these exercises **4 ROUNDS**

1 - High Plank: 25 sec.

2 - Chair: 20 sec.
Note: sit on an imaginary chair, keep your back straight.

3 - Waist Hooping: 20 times.
Note: if you do not have a hoop, pretend you have an imaginary hoop and rotate your hips 10 times.

4 - Abs: 25 times

MATH

Geometry

1. Find the missing angle below.

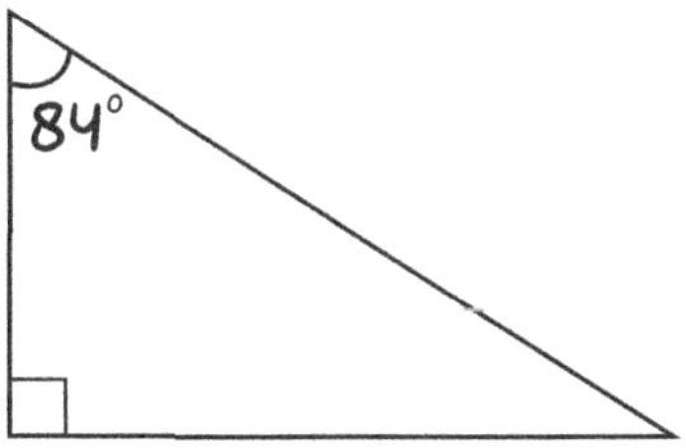

A. 90°　　**C.** 4°
B. 2°　　**D.** 6°

2. Find the missing angle below.

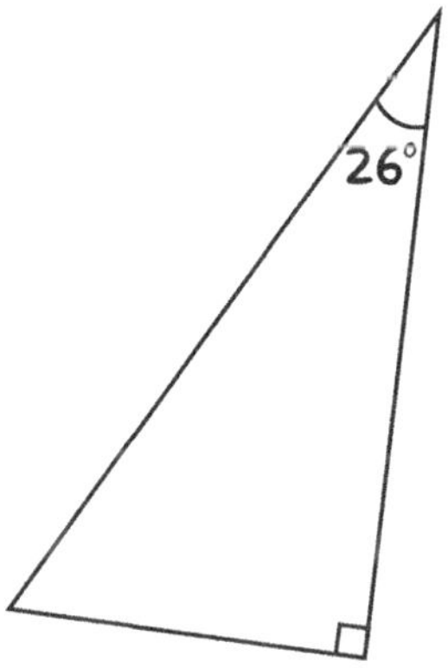

A. 64°　　**C.** 68°
B. 66°　　**D.** 70°

3. Find the missing angle below.

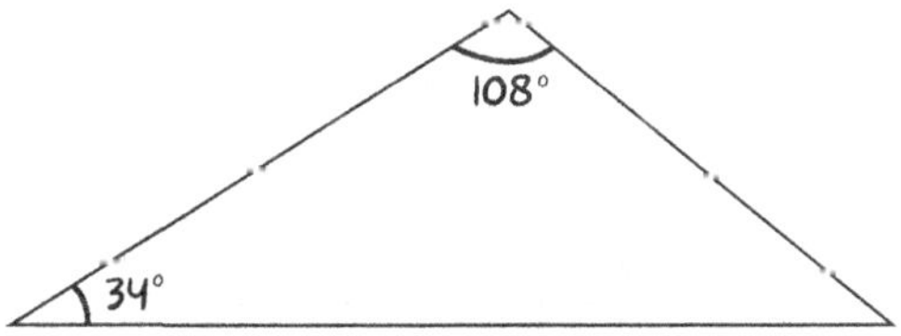

A. 38°　　**C.** 42°
B. 40°　　**D.** 44°

4. Find the missing angle below.

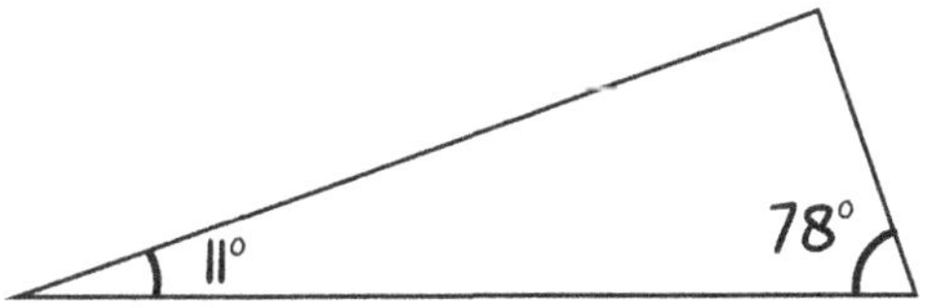

A. 87°　　**C.** 91°
B. 89°　　**D.** 93°

5. Find the missing angle below.

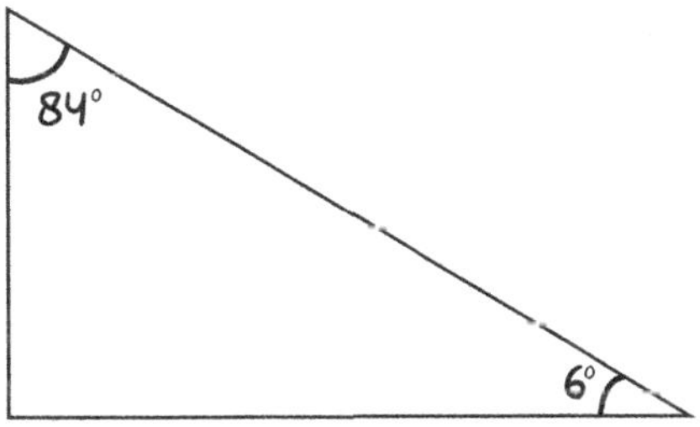

A. 86°　　**C.** 90°
B. 88°　　**D.** 92°

6. Find the missing angle below.

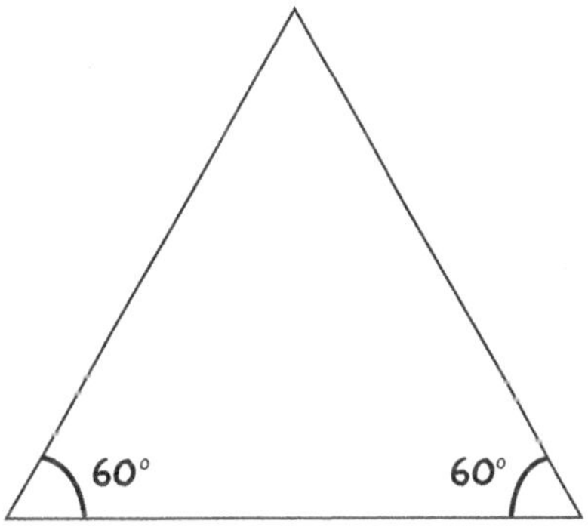

A. 35°　　**C.** 45°
B. 50°　　**D.** 60°

WEEK 8 DAY 5 MATH

Geometry

7. Find the missing angle below.

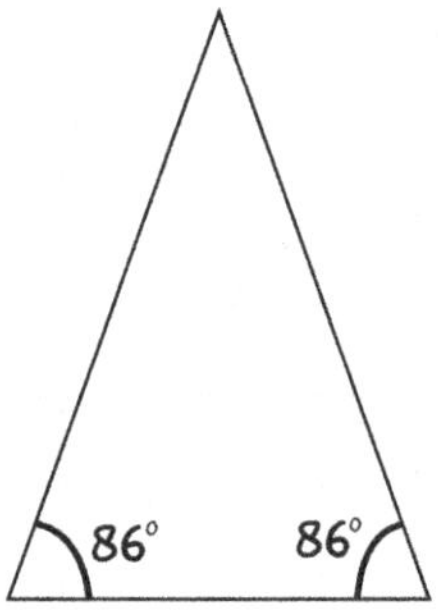

A. 8° **C.** 12°
B. 10° **D.** 14°

8. Find the missing angle below.

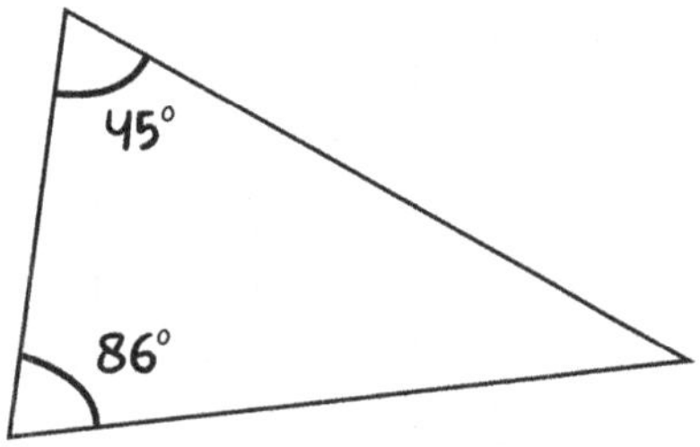

A. 43° **C.** 47°
B. 45° **D.** 49°

9. Find the missing angle below.

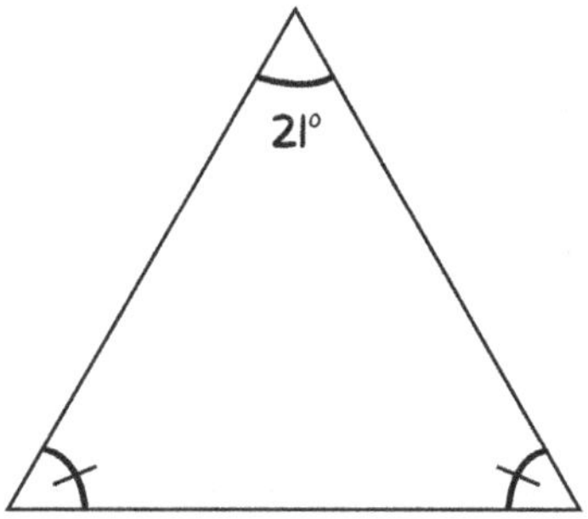

A. 79.5° **C.** 79°
B. 78° **D.** 159°

10. Find the missing angle below.

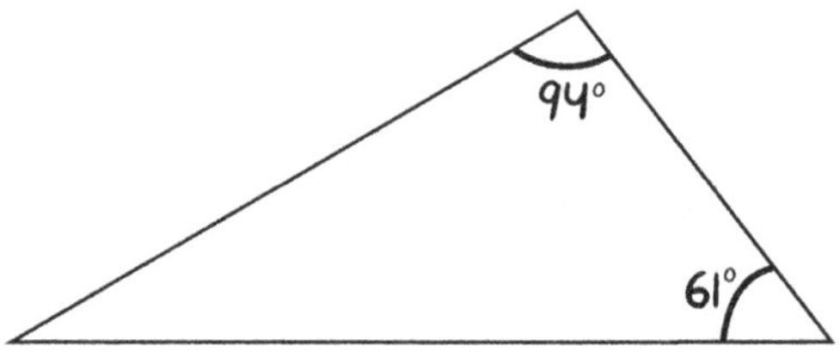

A. 20° **C.** 30°
B. 25° **D.** 35°

YOGA

Please be aware of your environment and be safe at all times. If you cannot do an exercise, just try your best.

1 - Down Dog: 30 sec.

2 - Bend Down: 30 sec.

3 - Chair: 30 sec.

4 - Child Pose: 30 sec.

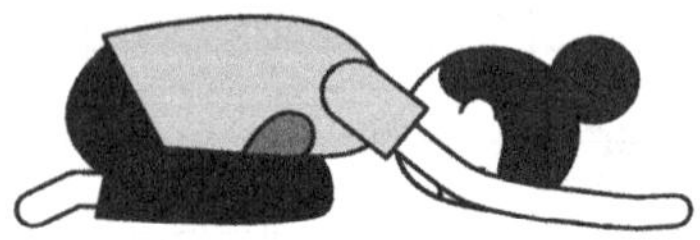

5 - Shavasana: as long as you can. Note: think of happy moments and relax your mind.

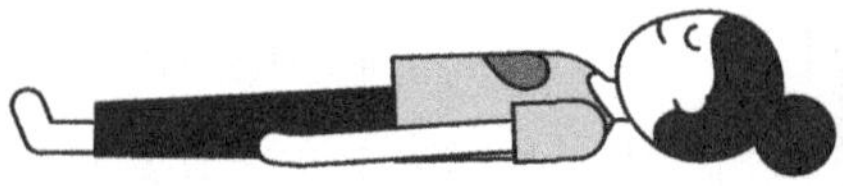

WEEK 8 DAY 6

EXPERIMENT

Cellular Reproduction - Mitosis

You probably already know that all living organisms are made up of ***cells****. The simplest plants, animals, and fungi may be made of a* ***single cell****, but most of the organisms that we can see or interact with are made up of billions or trillions of cells. However, like all living things, cells have a* ***lifespan****. Since they do not last forever, cells must* ***reproduce*** *to create new cells that can continue their work once they are gone. New cells are also needed so that plants and animals can* ***grow*** *- think of how many more cells are in your body now than when you were a tiny baby!*

Most *cells reproduce through a process known as* ***mitosis****, sometimes commonly referred to as "cell division." The word "****division****" is used because, during* ***mitosis****, the cell literally divides itself into two different cells. Today, we'll do an activity that will make that process much more visual and easier to understand!*

Materials:

- 1 large poster board
- A ruler or yardstick
- Art supplies (markers, colored pencils, etc.)
- At least 2 different colors of yarn
- Scissors
- Glue
- A compass (the kind that draws circles) - optional
- An encyclopedia and/or Internet Access (for research)
- Notepaper

WEEK 8 DAY 6 EXPERIMENT

Procedure:

1. At the top of your poster board, write the title "MITOSIS."

2. Using your ruler and markers, divide the remainder of the poster board into **eight** equal sized boxes. Label your boxes from 1-8 in this manner: **Prophase I, Metaphase I, Anaphase I, Telophase I, Prophase II, Metaphase II, Anaphase II, Telophase II.**

3. Using the internet (or encyclopedias) for research, look up the process of **mitosis** and the **eight phases** mentioned above. On your notepaper, write down **what's happening** in each of these 8 phases of cell division. Try to understand both what the process "looks like" and what is happening inside the cell as it reproduces itself.

4. Once you feel like you have a general understanding of all 8 phases, use your art supplies to recreate the 8 steps of mitosis on your poster board. Use your art supplies to create the cells and cut pieces of yarn to represent the **chromosomes** that are being copied and moved throughout the process. Be sure to refer to the notes you created in Step 3.

5. Once your poster board is completed, clean up your materials and answer the questions below. Hold onto your mitosis poster, as you'll compare it to a different cell reproduction process next week!

WEEK 8 DAY 6 EXPERIMENT

Follow-Up Questions:

1. Based on the different drawings you made and pieces of yarn you placed, which step in the process of **mitosis** do you think is the most **complex**, and why?

2. Based on what you saw today, what kind of "**mistakes**" do you think could happen during mitosis that could result in a "bad" cell?

YOGA

Please be aware of your environment and be safe at all times. If you cannot do an exercise, just try your best.

2 - Down Dog: 30 sec.

3 - Stretching: Stay as long as possible.
Note: do on one leg then on another.

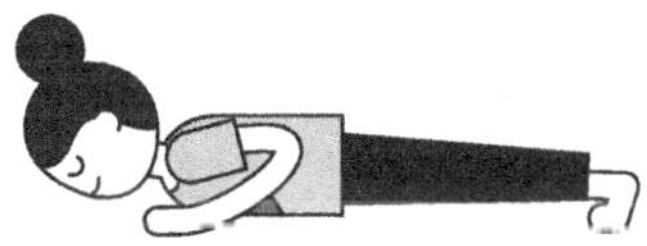

4 - Lower Plank: 20 sec.
Note: Keep your back straight and body tight.

1 - Tree Pose: Stay as long as possible.
Note: do on one leg then on another.

6 - Shavasana: 5 min.
Note: this pose is very important and provides you with long term benefits. Try not to skip this. Close your eyes and imagine who you want to be and what your goals are! Always think happy thoughts.

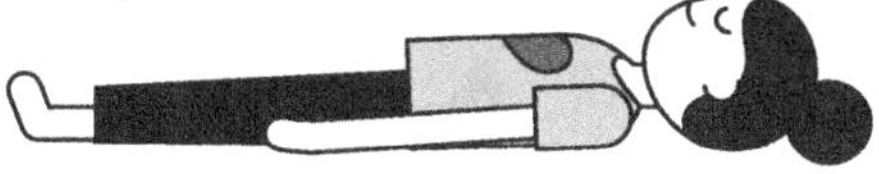

5 -Book Pose: 20 sec.
Note: Keep your core tight. Legs should be across from your eyes.

LOGIC

Task:

Take a look at the first row (numbers labeled from 1 to 6). The following rows must have the same pictures but the order does not matter. Figure out the missing pictures for each row.

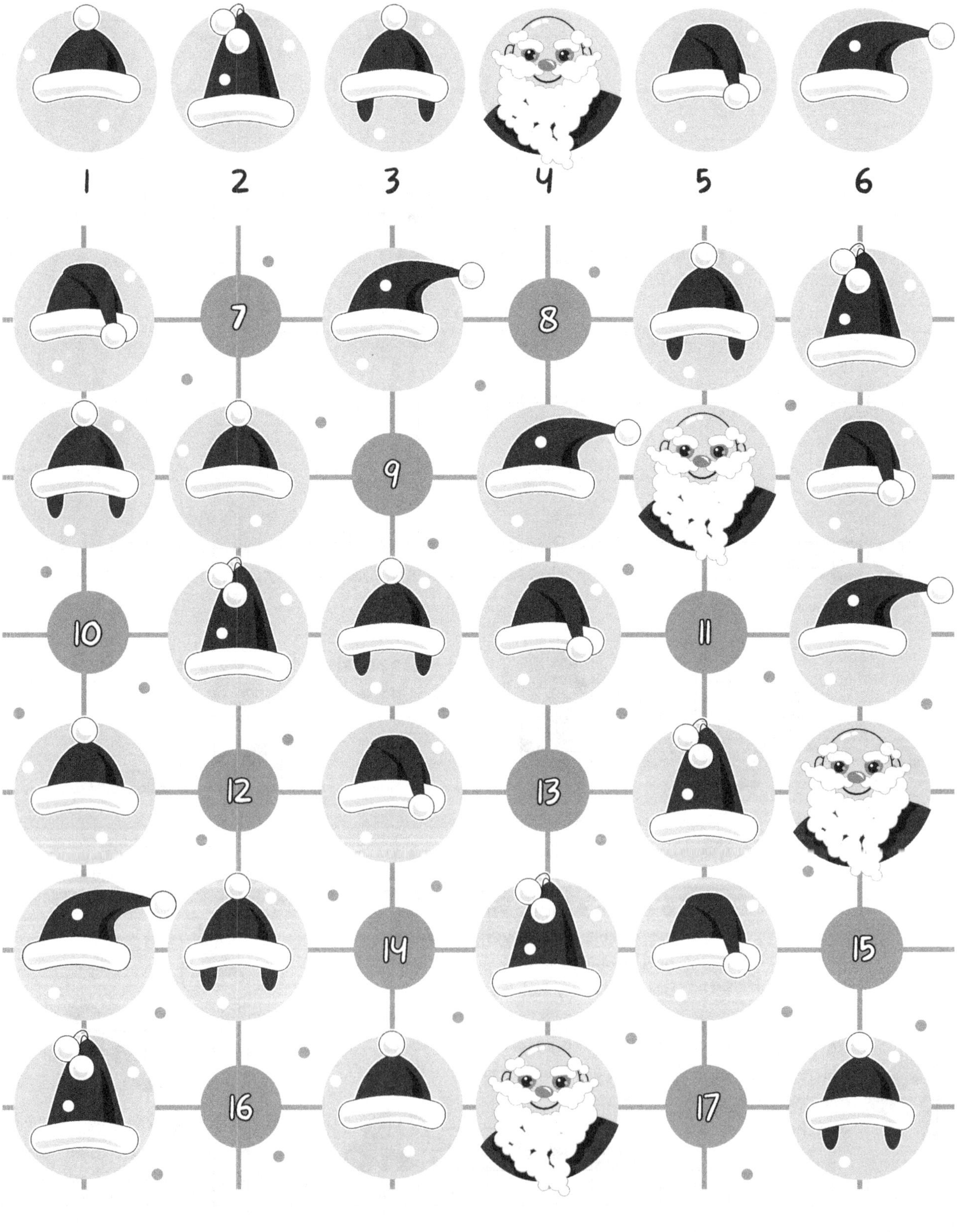

WEEK 9

Rock on!
When working on word problems in the math section, always underline keywords and make sure to double check the answer to see if it makes sense.
POLICE

WEEK 9 DAY 1 OVERVIEW OF ENGLISH CONCEPTS

Relationships Between Ideas in Informational Text

Last week, we looked at the concept of **structure** in informational text and saw that creating clear sections (an intro, body, and conclusion) and embracing organization makes both reading and writing much simpler. This week, we'll extend that thinking and explore what effective organization can achieve: clear relationships between ideas.

Sometimes when we read, we have a tendency to notice ideas individually. When your teacher asks you "What three main ideas did this text explore?" you can probably answer that question with no problem, as long as you read closely. However, when the teacher asks, "How are those three themes connected or interrelated?" it's a **much** higher-order question. Determining and analyzing the relationships between ideas requires a slightly new set of strategies and tools!

Preparing to Connect:

As you read, you should pay close attention (and hopefully take notes) when new ideas are presented. Just paying attention isn't enough to make connections, though! Making connections requires...

- **Strong Comprehension of the Text** – If you're not sure what the text is all about, it's going to be incredibly difficult to make connections. Before you embrace these high-order questions, you need to build a solid understanding of the **literal**, **explicit** meaning of the text.
- **Time and Patience** – When you have to connect ideas, it's impossible to simply check the right page in the book and find the "correct answer." Thinking deep and connecting ideas in big ways requires that you sit down and devote some time to thinking about a text
- **An Open Mind** – Remember, there might be more to the text than you noticed on the first reading
- **Other Inquisitive Minds** – Making connections by yourself can be tough or lonely work. The more brains involved, the more different connections people are likely to make. If you're having trouble thinking about two texts together or thinking about a variety of different places in the same text, call upon some friends who are reading the same things. Together, you can push each other's thinking and get more results!

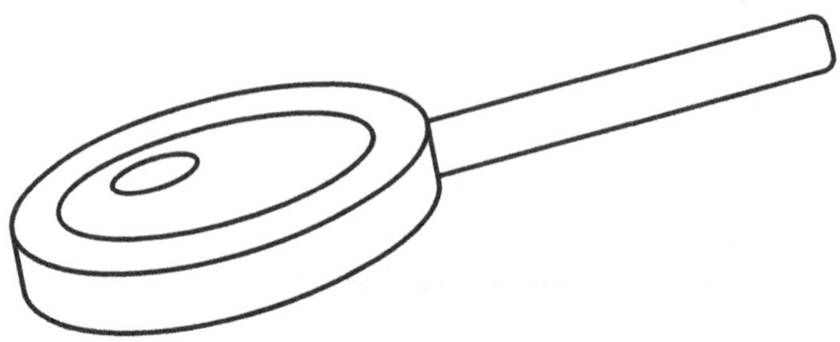

WEEK 9 DAY 1

OVERVIEW OF ENGLISH CONCEPTS

Connecting Different Ideas:

As you look to make different connections as a reader, always think about...

- How ideas introduced **early** in the text affect things that happen later
- How individual, small ideas or concepts introduced early **build up or come together to become bigger ideas** or major concepts later in the text
- Which main ideas or points does the author **keep coming back to**?
    - How are these ideas, points, or principles applicable in a variety of situations throughout the text?
- How is the author trying to guide your thinking? How are they trying to **teach** you new things?
    - This often gets back to the idea of **organization**! If you understand why the author organized the text the way they did, then you can make better conclusions about **what relationships between ideas** they want you to focus on.

Trusting Yourself as a Reader & Writer:

- Making connections as a reader requires some courage because you could wind up being wrong or misunderstanding things. That's okay! The most important thing is that you're tackling complex texts head on and trying to understand them fully!
- As a writer, you must always strive to write as clearly and effectively as possible. That means reflecting on the best way to explain things and using examples that make the connections between ideas clear. If readers come away unsure of how ideas are related, that's your problem! Getting it right requires a lot of practice, so be patient with yourself as you work toward that goal!

From "The Narrative of the Life of Frederick Douglass, An American Slave"

By Frederick Douglass

My master's family consisted of two sons, Andrew and Richard; one daughter, Lucretia, and her husband, Captain Thomas Auld. They lived in one house, upon the home plantation of Colonel Edward Lloyd. My master was Colonel Lloyd's clerk and superintendent. He was what might be called the overseer of the overseers. I spent two years of childhood on this plantation in my old master's family. It was here I received my first impressions of slavery on this plantation, so I will give some description of it, and of slavery as it there existed. The plantation is about twelve miles north of Easton, in Talbot county, and is situated on the border of Miles River. The principal products raised upon it were tobacco, corn, and wheat. These were raised in great abundance; so that, with the products of this and the other farms belonging to him, he was able to keep in almost constant employment a large sloop, in carrying them to market at Baltimore. This sloop was named Sally Lloyd, in honor of one of the colonel's daughters. My master's son-in-law, Captain Auld, was master of the vessel; she was otherwise manned by the colonel's own slaves. Their names were Peter, Isaac, Rich, and Jake. These were esteemed very highly by the other slaves, and looked upon as the privileged ones of the plantation; for it was no small affair, in the eyes of the slaves, to be allowed to see Baltimore.

Colonel Lloyd kept from three to four hundred slaves on his home plantation, and owned a large number more on the neighboring farms belonging to him. The names of the farms nearest to the home plantation were Wye Town and New Design. "Wye Town" was under the overseership of a man named Noah Willis. New Design was under the overseership of a Mr. Townsend. The overseers of these, and all the rest of the farms, numbering over twenty, received advice and direction from the managers of the home plantation. This was the great business place. It was the seat of government for the whole twenty farms. All disputes among the overseers were settled here. If a slave was convicted of any high misdemeanor, became unmanageable, or evinced a determination to run away, he was brought immediately here, severely whipped, put on board the sloop, carried to Baltimore, and sold to Austin Woolfolk, or some other slave-trader, as a warning to the slaves remaining.

Here, too, the slaves of all the other farms received their monthly allowance of food, and their yearly clothing. The men and women slaves received, as their monthly allowance of food, eight pounds of pork, or its equivalent in fish, and one bushel of corn meal. Their yearly clothing consisted of two coarse linen shirts, one pair of linen trousers, like the shirts, one jacket, one pair of trousers for winter, made of coarse cloth, one pair of stockings, and one pair of shoes; the whole of which could not have cost more than seven dollars. The allowance of the slave children was given to their mothers, or the old women having the care of them. The children unable to work in the field had neither shoes, stockings, jackets, nor trousers, given to them; their clothing consisted of two coarse linen shirts per year. When these failed them, they went naked until the next allowance-day. Children from seven to ten years old, of both sexes, almost naked, might be seen at all seasons of the year.

READING PASSAGE

1. Based on the passage, how do we know that **Colonel Lloyd** and his family are extremely successful?

2. Why do the other slaves consider **Peter, Isaac, Rich, and Jake** to be special or privileged?

3. Based on the passage, what is a **sloop**?

   A. A kind of slave
   B. A kind of farm
   C. A kind of shipment
   D. A kind of boat

4. How many farms does Colonel Lloyd control?

   A. One
   B. 300-400
   C. Three
   D. 20-30

5. In what ways is the "allowance" given to slaves inadequate?

Thinking About Relationships Between Ideas (Pt. 1)

Directions: Today, you'll reflect back on the ***Narrative of Frederick Douglass*** passages that you've read so far to think about the relationships between some of the ideas in the text. You will probably want to flip back through this book to some earlier sections to review the texts again.

1. Back on **Week 8, Day 1**, Douglass began by talking about a main difference between how white children and black children grew up. How is this same idea reflected in **Week 9, Day 1's** text?

2. How has **transportation** (and lack thereof) been a major theme or repeated idea throughout the first passages of the narrative? Provide specific examples!

OVERVIEW OF ENGLISH CONCEPTS
ACTIVITIES

3. How has Douglass exposed the **problems of slavery** without specifically assigning blame or judgment to his former masters?

FITNESS

Please be aware of your environment and be safe at all times. If you cannot do an exercise, just try your best.

Repeat these exercises **4 ROUNDS**

2 - Lunges: 8 times to each leg.
Note: Use your body weight or books as weight to do leg lunges.

1 - Abs: 20 times

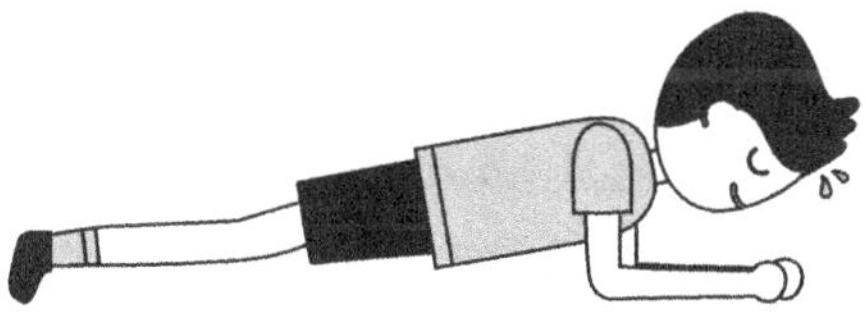

3 - Plank: 20 sec.

4 - Run: 50m
Note: Run **25** meters to one side and **25** meters back to the starting position.

From "The Narrative of the Life of Frederick Douglass, An American Slave"

By Frederick Douglass

(Continued from Day 1's Passage)
The home plantation of Colonel Lloyd wore the appearance of a country village. All the mechanical operations for all the farms were performed here. The shoemaking and mending, the blacksmithing, cartwrighting, coopering, weaving, and grain-grinding, were all performed by the slaves on the home plantation. The whole place wore a business-like aspect very unlike the neighboring farms. The number of houses, too, conspired to give it advantage over the neighboring farms. It was called by the slaves the Great House Farm. Few privileges were esteemed higher, by the slaves of the out-farms, than that of being selected to do errands at the Great House Farm. It was associated in their minds with greatness. A representative could not be prouder of his election to a seat in the American Congress, than a slave on one of the out-farms would be of his election to do errands at the Great House Farm. They regarded it as evidence of great confidence reposed in them by their overseers; and it was on this account, as well as a constant desire to be out of the field from under the driver's lash, that they esteemed it a high privilege, one worth careful living for. He was called the smartest and most trusty fellow, who had this honor conferred upon him the most frequently. The competitors for this office sought as diligently to please their overseers, as the office-seekers in the political parties seek to please and deceive the people. The same traits of character might be seen in Colonel Lloyd's slaves, as are seen in the slaves of the political parties.

The slaves selected to go to the Great House Farm, for the monthly allowance for themselves and their fellow-slaves, were peculiarly enthusiastic. While on their way, they would make the dense old woods, for miles around, reverberate with their wild songs, revealing at once the highest joy and the deepest sadness. They would compose and sing as they went along, consulting neither time nor tune. The thought that came up, came out — if not in the word, in the sound; — and as frequently in the one as in the other. They would sometimes sing the most pathetic sentiment in the most rapturous tone, and the most rapturous sentiment in the most pathetic tone. Into all of their songs they would manage to weave something of the Great House Farm. Especially would they do this, when leaving home. They would then sing most exultingly the following words: —

"I am going away to the Great House Farm!

O, yea! O, yea! O!"

This they would sing, as a chorus, to words which to many would seem unmeaning jargon, but which, nevertheless, were full of meaning to themselves. I have sometimes thought that the mere hearing of those songs would do more to impress some minds with the horrible character of slavery, than the reading of whole volumes of philosophy on the subject could do.

OVERVIEW OF ENGLISH CONCEPTS
READING PASSAGE

1. How is the "**home plantation**" different from the other farms?

2. Why do the slaves refer to the home plantation as "**the Great Farm House**"?

3. Why does Douglass compare the **plantation** to the world of **politics**?

   A. In both situations, people work without pay in bad conditions.
   B. In both situations, people lie and deceive others to better their own position.
   C. In both situations, people can vote so their voice is heard.
   D. In both situations, people work hard to be recognized as the best.

4. How do slaves react when they're sent to the **Great Farm House**?

   A. They celebrate.
   B. They cry.
   C. They fear for their lives.
   D. They are disappointed.

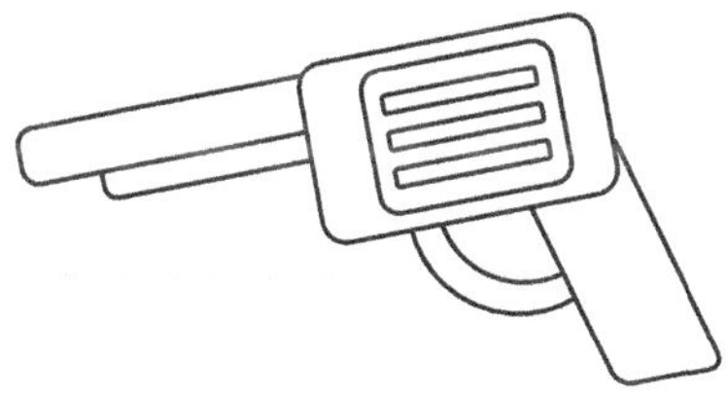

5. How do the slaves' **songs** as they travel to the Great Farm House reveal the complexity and sadness of the whole situation?

Thinking About Relationships Between Ideas (Pt. 2)

Directions: Today, you'll reflect back on the ***Narrative of Frederick Douglass*** passages that you've read so far to think about the relationships between some of the ideas in the text. You will probably want to flip back through this book to some earlier sections to review the texts again.

1. How have **selfishness and greed** been major themes or repeated ideas throughout the first passages of the narrative? Provide specific examples!

2. How does the way the slaves **celebrate** when they find out they have been sent to the Great Farm House **differ** from other portrayals of slave life we've seen in previous passages?

OVERVIEW OF ENGLISH CONCEPTS
ACTIVITIES

3. Based on what you've read so far, what aspects of slavery and his upbringing did Frederick Douglass think **were most important for people to know about and understand**? How does he get those ideas across in the text?

____

____

____

____

____

____

FITNESS

Please be aware of your environment and be safe at all times. If you cannot do an exercise, just try your best.

Repeat these exercises **4 ROUNDS**

1 - Squats: 20 times.
Note: imagine you are trying to sit on a chair.

2 - Side Bending: 15 times to each side. Note: try to touch your feet.

3 - Tree Pose: Stay as long as possible.
Note: do the same with the other leg.

MATH

Statistics

1. Amelia surveyed 80 students at her school about their favorite subject. Of those surveyed, 20 chose science as their favorite subject. If there are 600 students in the whole school, around how many like science?

   A. 200 C. 300
   B. 80 D. 150

2. Amelia surveyed 80 students at her school about their favorite subject. Of those surveyed, 5 chose math as their favorite subject. If there are 750 students in the whole school, around how many like math?

   A. 45 C. 80
   B. 90 D. 500

3. Alex is trying to construct a random sample to make inferences about how many students have siblings. If the total population is 300, which amount would be an acceptable random sample?

   A. 3 C. 45
   B. 13 D. 200

4. A group of randomly selected shoppers were asked what type of bag they prefer to use. If 20 shoppers chose cloth bags out of 60, what percentage of shoppers prefer cloth bags?

   A. 20% C. 66%
   B. 33% D. 90%

5. A doctor's office survey asked a random sampling of patients about their exercise habits. If they have 230 patients and they estimate 115 exercise regularly, how many people out of a survey of 60 people said they exercise regularly?

   A. 10 C. 15
   B. 20 D. 30

6. You want to figure out if students will recycle after lunch in your school. What would be the best group to use as a random sample?

   A. 20 teachers
   B. Administrators
   C. 20 random people
   D. 20 random students in the lunchroom

7. You want to know what students' favorite movie snack is. What would be the best group to use as a random sample?

   A. The first 20 kids you see at the movies
   B. The first 5 kids you see at the movies
   C. The first 200 kids you see at the movies
   D. The first 20 people you see at the movies

8. You want to know which cartoon is most popular with 8 year olds. What would be the best random sample?

   A. 50 10-year olds
   B. 50 8-year olds
   C. 50 6-year olds
   D. 50 kids of random ages

WEEK 9 DAY 3 MATH

9. You want to know what song is most popular with middle schoolers. What would be the best group to use as a random sample?

   A. 20 students from the band
   B. 20 students from the orchestra
   C. 20 teachers
   D. 20 students from 3 random homerooms

10. You want to know the most popular snack. What would be the best group to use as a random sample?

    A. 30 people in the supermarket snack aisle
    B. 300 people in the supermarket snack aisle
    C. 3 people in the supermarket snack aisle
    D. 3,000 people in the supermarket snack aisle

11. Out of 2000 cars produced, the manufacturer took a random sample of 50 cars. Of those 50, 2 had a defect. Estimate the total number of defective cars.

    A. 65 C. 105
    B. 80 D. 405

12. From 7500 TVs produced, the manufacturer took a random sample of 30 TVs. Of those 30, one had a defect. Estimate the total number of defective TVs.

    A. 200 C. 300
    B. 250 D. 350

13. From 10,000 computers produced, the manufacturer took a random sample of 100 computers. Of those 100, 17 had a defect. Estimate the total number of defective computers.

    A. 17 C. 1700
    B. 170 D. 17,000

14. A farm worker inspected 200 apples from a barrel of 5000. She finds 35 apples are bad. How many apples should she estimate are bad in the total barrel?

    A. 800 C. 875
    B. 825 D. 925

FITNESS

Please be aware of your environment and be safe at all times. If you cannot do an exercise, just try your best.

Repeat these exercises **4 ROUNDS**

1 - Bend forward: 15 times.
Note: try to touch your feet. Make sure to keep your back straight and if needed you can bend your knees.

2 - Lunges: 12 times to each leg.
Note: Use your body weight or books as weight to do leg lunges.

3 - Plank: 20 sec.

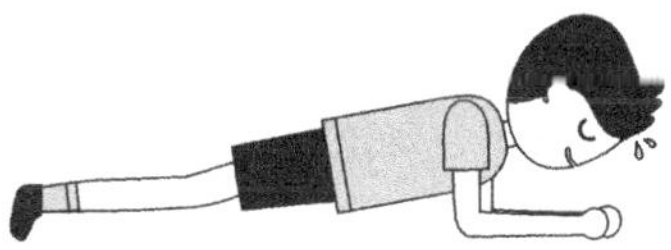

4 - Abs: 20 times

WEEK 9 DAY 4 MATH

Statistics

1. You want to know how popular basketball is at your school. Which group would give you the best random sample?

- **A.** A selection of students from mandatory gym class
- **B.** A selection of the tallest students
- **C.** A selection of the smallest students
- **D.** A selection of students from the basketball team

2. You want to know which restaurant in your town is most popular. Which group would give you the best random sample?

- **A.** A selection of people at the local mall
- **B.** A selection of people at an expensive restaurant
- **C.** A selection of teens in your neighborhood
- **D.** A selection of elderly people from your church

3. You want to know how many elementary students bring their lunch to school. Which group would give you the best random sample?

- **A.** 30 1st graders
- **B.** 30 6th graders
- **C.** 30 10th graders
- **D.** 30 elementary parents

4. You want to know how many students your age have to do chores at home. Which group would give you the best random sample?

- **A.** 4 students in your grade
- **B.** 40 students in your grade
- **C.** 400 students in your grade
- **D.** 4000 students from your town

5. You want to know how many kids like to go to the beach on vacation. Which group would give you the best random sample?

- **A.** 20 kids at a park
- **B.** 20 kids on the beach
- **C.** 20 kids from your school
- **D.** 20 kids hiking in the woods

6. According to the results of last year's standardized tests, 20% of 6th graders scored above grade level in math. What is a valid conclusion to make?

- **A.** If there are 100 6th grade students, 40 are above grade level.
- **B.** If there are 400 6th grade students, 40 are above grade level.
- **C.** If there are 200 6th grade students, 40 are above grade level.
- **D.** If there are 300 6th grade students, 40 are above grade level.

7. According to the results of last year's standardized tests, 40% of 8th graders scored below grade level in math. What is a valid conclusion to make?

- **A.** If there are 150 8th grade students, 120 are below grade level.
- **B.** If there are 150 8th grade students, 60 are below grade level.
- **C.** If there are 150 8th grade students, 80 are below grade level.
- **D.** If there are 150 8th grade students, 50 are below grade level.

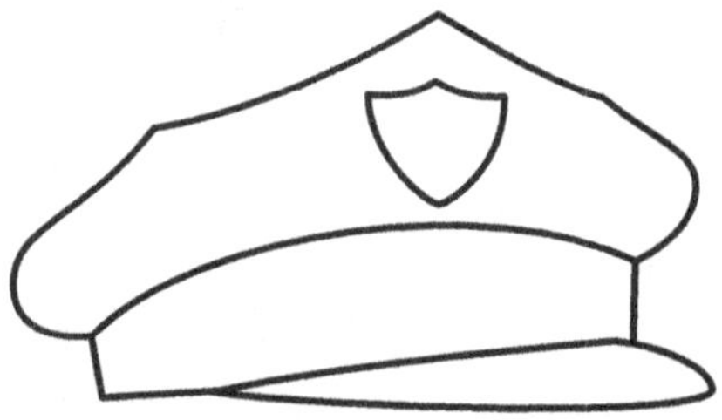

WEEK 9 DAY 4

MATH

8. According to the results of last years standardized tests, 30% of 7th graders scored at grade level in math. What is a valid conclusion to make?

   A. If there are 200 students, 60 are above grade level.
   B. If there are 200 students, 60 are below grade level.
   C. If there are 200 students, 90 are at grade level.
   D. If there are 200 students, 60 are at grade level.

9. A company wants to add a new flavor of ice cream. They conduct sample tests with three flavors. The data showed 65% liked mango, 20% liked dragonfruit and 10% did not like either. What is a valid conclusion to make?

   A. The company should add mango as their new flavor.
   B. The company should add dragon-fruit as their new flavor.
   C. The company should do more research.
   D. The company should not add either flavor.

10. A company wants to add a new flavor of ice cream. They conduct sample tests with three flavors. No clear flavor had a majority of people who enjoyed it. What is a valid conclusion to make?

    A. The company should do more research.
    B. The company should add all three flavors.
    C. The company should try three other flavors.
    D. The company should pick one flavor to add.

11. You want to know what students' favorite sport is. Which group would give you the best random sample?

    A. A group of students on your baseball team
    B. A group of students who run track
    C. A group of students from your gym class
    D. A group of students who go to all the football games

FITNESS

Please be aware of your environment and be safe at all times. If you cannot do an exercise, just try your best.

Repeat these exercises **4 ROUNDS**

1 - High Plank: 25 sec.

2 - Chair: 20 sec.
Note: sit on an imaginary chair, keep your back straight.

3 - Waist Hooping: 20 times.
Note: if you do not have a hoop, pretend you have an imaginary hoop and rotate your hips 10 times.

4 - Abs: 25 times

MATH

Statistics

1. You toss three coins. Create a tree diagram to show your sample space.

2. You toss three coins. Find P (3 heads)

A. $\frac{1}{8}$
B. $\frac{1}{9}$
C. $\frac{1}{3}$
D. $\frac{6}{9}$

3. You toss three coins. Find P (two tails)

A. $\frac{2}{8}$
B. $\frac{3}{8}$
C. $\frac{6}{8}$
D. $\frac{7}{8}$

4. You toss three coins. Find P (at least one head)

A. $\frac{4}{8}$
B. $\frac{6}{8}$
C. $\frac{5}{8}$
D. $\frac{7}{8}$

5. You toss three coins. Find P (at least one tail)

A. $\frac{8}{8}$
B. $\frac{6}{8}$
C. $\frac{7}{8}$
D. $\frac{4}{8}$

6. You have 2 number cubes. Create a tree diagram to show your sample space.

7. You have 2 number cubes. Find P(same number twice)

A. $\frac{6}{30}$
B. $\frac{6}{36}$
C. $\frac{10}{36}$
D. $\frac{15}{30}$

8. You have 2 number cubes. Find P(1 and 6)

A. $\frac{2}{36}$
B. $\frac{2}{18}$
C. $\frac{4}{18}$
D. $\frac{2}{9}$

9. You have 2 number cubes. Find P(2 and 2)

A. $\frac{1}{6}$
B. $\frac{1}{10}$
C. $\frac{1}{24}$
D. $\frac{1}{36}$

10. You have 2 number cubes. Find P(3 and 4)

A. $\frac{4}{36}$
B. $\frac{3}{36}$
C. $\frac{1}{36}$
D. $\frac{2}{36}$

MATH

11. You have 2 number cubes. Find P(sum of cubes is 12)

A. $\frac{1}{36}$　　C. $\frac{3}{36}$

B. $\frac{2}{36}$　　D. $\frac{4}{36}$

12. You choose a shirt and a pair of pants from the following color choices: Shirt: blue, green, gray, purple, white; Pants: blue, khaki, black. Create a tree diagram with your sample space.

13. You choose a shirt and a pair of pants. P(same color shirt and pants)

A. $\frac{1}{5}$　　C. $\frac{1}{3}$

B. $\frac{1}{4}$　　D. $\frac{1}{15}$

14. You choose a shirt and a pair of pants. P(khaki pants)

A. $\frac{1}{5}$　　C. $\frac{1}{4}$

B. $\frac{1}{3}$　　D. $\frac{1}{15}$

15. You choose a shirt and a pair of pants. P(green shirt)

A. $\frac{1}{5}$　　C. $\frac{1}{2}$

B. $\frac{1}{3}$　　D. $\frac{1}{6}$

YOGA

Please be aware of your environment and be safe at all times. If you cannot do an exercise, just try your best.

1 - Down Dog: 30 sec.

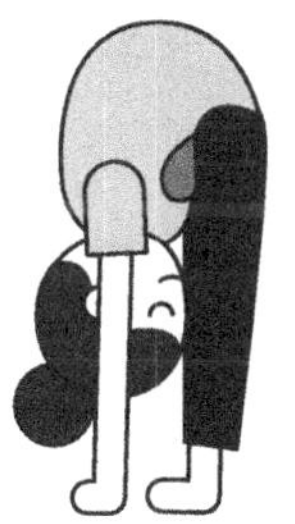

2 - Bend Down: 30 sec.

3 - Chair: 30 sec.

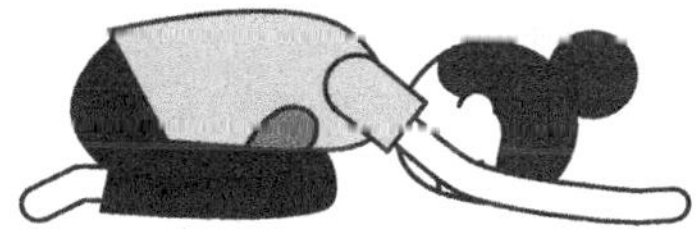

4 - Child Pose: 30 sec.

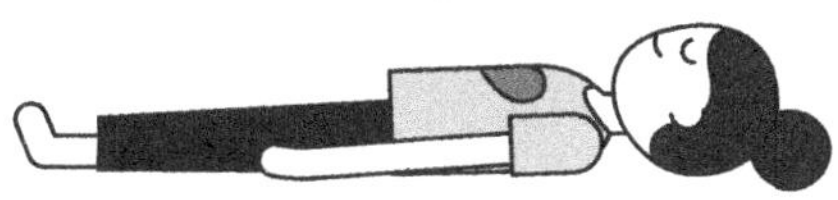

5 - Shavasana: as long as you can. Note: think of happy moments and relax your mind.

WEEK 9 DAY 6

EXPERIMENT

Cellular Reproduction: Meiosis

*Last week, we introduced the idea of cellular reproduction through a process known as **mitosis**. In **mitosis**, one cell copies its genetic information and splits into two cells, a process that can be repeated over and over. While **most** cells reproduce through mitosis, some very special and important cells reproduce through a slightly different process known as **meiosis**.*

***Meiosis** is how special reproductive cells, known as gametes, reproduce within our bodies. Those gametes, commonly known as **sperm** and **eggs** are intentionally created with **incomplete** genetic information, so that they can combine with a complimentary cell from another source to form a new genetic code and create a new organism. Today, you'll create a model that makes **meiosis** highly visual and will help you understand the distinction between this process and the more common **mitosis**.*

Materials:

- Your **Mitosis** poster from last week
- 1 large poster board
- A ruler or yardstick
- Art supplies (markers, colored pencils, etc.)
- At least 2 different colors of yarn
- Scissors
- Glue
- A compass (the kind that draws circles) - optional
- An encyclopedia and/or Internet Access (for research)
- Notepaper

WEEK 9 DAY 6 EXPERIMENT

Procedure:

1. At the top of your poster board, write the title "MEIOSIS."

2. Using your ruler and markers, divide the remainder of the poster board into **eight** equal sized boxes. Label your boxes from 1-8 in this manner: **Prophase I, Metaphase I, Anaphase I, Telophase I, Prophase II, Metaphase II, Anaphase II, Telophase II**. If you want, you can label the first four together as "Meiosis I" and the second four together as "Meiosis II."

3. Using the internet (or encyclopedias) for research, look up the process of **meiosis** and the **eight phases** mentioned above. On your notepaper, write down **what's happening** in each of these 8 phases of cell division. Try to understand both what the process "looks like" and what is happening inside the cell as it reproduces itself.

4. Once you feel like you have a general understanding of all 8 phases, use your art supplies to recreate the 8 steps of meiosis on your poster board. Use your art supplies to create the cells and cut pieces of yarn to represent the **chromatids** that are being copied and moved throughout the process. Be sure to refer to the notes you created in Step 3.

5. Once your poster board is complete, grab your mitosis poster from last week and compare the two side-by-side to notice the similarities and differences between the two processes.

6. Clean up your materials and answer the questions below.

WEEK 9 DAY 6 EXPERIMENT

Follow-Up Questions:

1. Based on your posters, name two major **similarities** between the processes of mitosis and meiosis.

____

____

____

2. Based on your posters, name two major **differences** between the processes of mitosis and meiosis.

____

____

____

YOGA

Please be aware of your environment and be safe at all times. If you cannot do an exercise, just try your best.

2 - Down Dog: 30 sec.

3 - Stretching: Stay as long as possible. Note: do on one leg then on another.

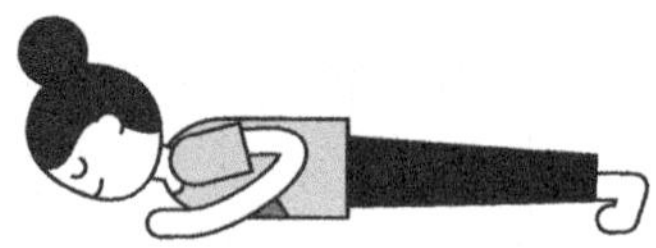

4 - Lower Plank: 20 sec. Note: Keep your back straight and body tight.

1 - Tree Pose: Stay as long as possible. Note: do on one leg then on another.

6 - Shavasana: 5 min. Note: this pose is very important and provides you with long term benefits. Try not to skip this. Close your eyes and imagine who you want to be and what your goals are! Always think happy thoughts.

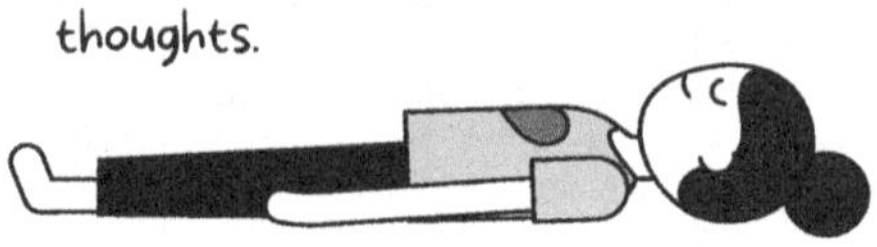

5 -Book Pose: 20 sec. Note: Keep your core tight. Legs should be across from your eyes.

MAZE

Task:

Find the way from the start to the end.

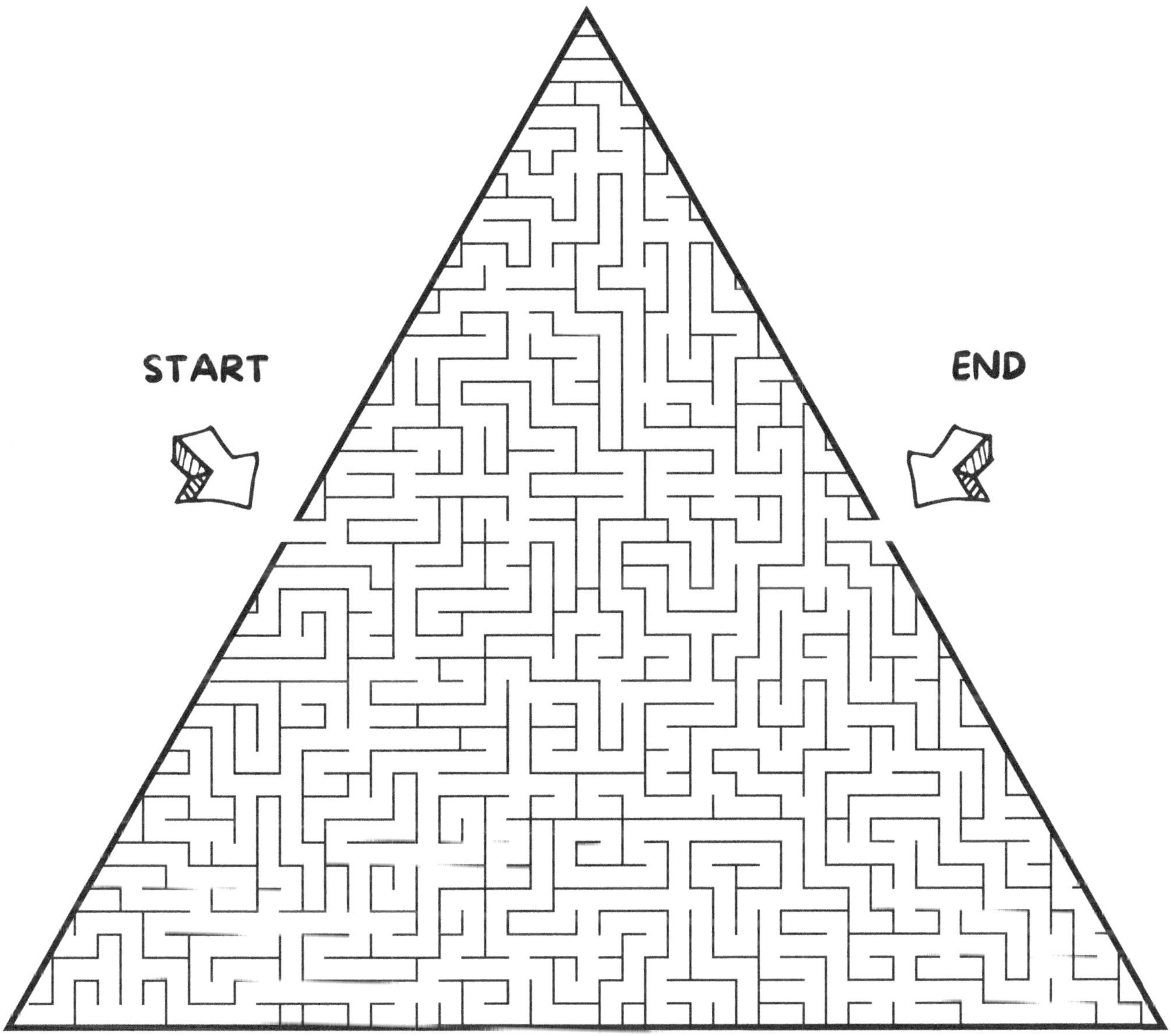

WEEK 10

ARGOPREP

OVERVIEW OF ENGLISH CONCEPTS

Irrelevant & Distracting Evidence

By now, you know that the goal of **informational writing** is to present information in as clear a way as possible to make ideas accessible and easy to understand for readers. Part of that work as a writer is making sure that you're only presenting readers with the **best, most useful,** and **most relevant** information. If something is off-topic or potentially distracting to the reader, you need to get it out of there!

With that said, sometimes authors will **intentionally** include irrelevant or distracting evidence when they write. Sometimes, this is done to confuse the reader and try to trick them into thinking or believing something so that they'll agree with the writer's viewpoint more readily. Other times, this distracting or irrelevant evidence is left in the text as a sort of "test" to make sure the reader knows what's really important to focus on. Sometimes, it's just a mistake! Regardless of the author's intent, you need to recognize irrelevant and distracting evidence as a reader!

Assessing Evidence:

When an author presents you with **evidence** (proof or support that backs up what they're saying), you need to **assess the strength and validity** of that material. That sounds like a lot of work, but you can actually achieve it by following these simple steps:

1. Read the text and underline or highlight examples of **evidence** the author is presenting
    a. Remember, these could be **quotes** from other texts, **statistics**, **examples** that illustrate ideas, **anecdotes** that help emphasize a point, etc.

2. Once you've identified the author's evidence/support, read each piece of evidence again on its own and ask yourself the following questions...
    a. Is this piece of evidence **directly connected to one of the author's main points or ideas?**
        i. If **not**, it may be irrelevant or distracting!
    b. What is the author **trying to achieve** by including this evidence here?
    c. Is this evidence **successfully achieving** what the author intended for it?
        i. If not, what's the problem? Is the evidence **irrelevant**? Should the author have chosen **something different**?
    d. Can I **verify this evidence** to ensure it's being presented and interpreted correctly?
        i. Authors **should** provide citations that make it easy for you to figure out where their evidence came from and double-check it
            1. If they're not citing sources, that's very **suspicious!**
            2. Go back and read the evidence **in its original context** to make sure the author is representing its actual meaning and not just cherry-picking facts!

3. Once you know for sure that a piece of evidence is **connected clearly to the text** and **properly represents the information that's being cited or quoted**, the only other step is to think about the points and the evidence together and check one more time that **everything makes sense!**

OVERVIEW OF ENGLISH CONCEPTS

From "Narrative of the Life of Frederick Douglass, An American Slave"

By Frederick Douglass

Colonel Lloyd kept a large and finely cultivated garden, which afforded almost constant employment for four men, besides the chief gardener, (Mr. M'Durmond.) This garden was probably the greatest attraction of the place. During the summer months, people came from far and near — from Baltimore, Easton, and Annapolis — to see it. It abounded in fruits of almost every description, from the hardy apple of the north to the delicate orange of the south. This garden was not the least source of trouble on the plantation. Its excellent fruit was quite a temptation to the hungry swarms of boys, as well as the older slaves, belonging to the colonel, few of whom had the virtue or the vice to resist it. Scarcely a day passed, during the summer, but that some slave had to take the lash for stealing fruit. The colonel had to resort to all kinds of stratagems to keep his slaves out of the garden. The last and most successful one was that of tarring his fence all around; after which, if a slave was caught with any tar upon his person, it was deemed sufficient proof that he had either been into the garden, or had tried to get in. In either case, he was severely whipped by the chief gardener. This plan worked well; the slaves became as fearful of tar as of the lash. They seemed to realize the impossibility of touching tar without being defiled.

The colonel also kept a splendid riding equipage. His stable and carriage-house presented the appearance of some of our large city livery establishments. His horses were of the finest form and noblest blood. His carriage-house contained three splendid coaches, three or four gigs, besides dearborns and barouches of the most fashionable style.

This establishment was under the care of two slaves — old Barney and young Barney — father and son. To attend to this establishment was their sole work. But it was by no means an easy employment; for in nothing was Colonel Lloyd more particular than in the management of his horses. The slightest inattention to these was unpardonable, and was visited upon those, under whose care they were placed, with the severest punishment; no excuse could shield them, if the colonel only suspected any want of attention to his horses — a supposition which he frequently indulged, and one which, of course, made the office of old and young Barney a very trying one. They never knew when they were safe from punishment. They were frequently whipped when least deserving, and escaped whipping when most deserving it. Every thing depended upon the looks of the horses, and the state of Colonel Lloyd's own mind when his horses were brought to him for use. If a horse did not move fast enough, or hold his head high enough, it was owing to some fault of his keepers. It was painful to stand near the stable-door, and hear the various complaints against the keepers when a horse was taken out for use. "This horse has not had proper attention. He has not been sufficiently rubbed and curried, or he has not been properly fed; his food was too wet or too dry; he got it too soon or too late; he was too hot or too cold; he had too much hay, and not enough of grain; or he had too much grain, and not enough of hay; instead of old Barney's attending to the horse, he had very improperly left it to his son."

READING PASSAGE

1. What is special or remarkable about **Colonel Lloyd's garden**?

2. According to the text, why were slaves **with tar on their clothing** beaten?

3. Paragraph 2 describes Colonel Lloyd's impressive collection of...

   A. Slaves
   B. Guns
   C. Vehicles
   D. Crops

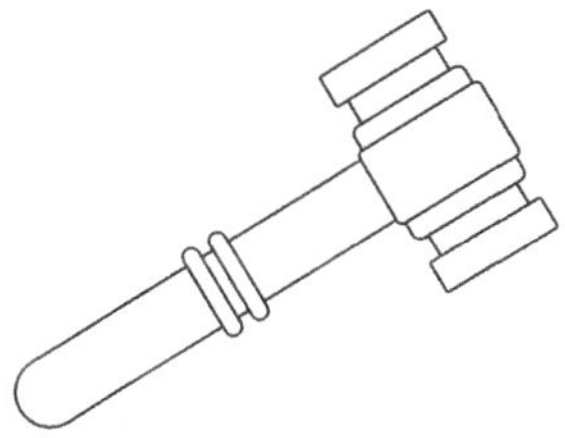

4. What does the phrase "**this establishment**" refer to at the beginning of Paragraph 3?

   A. Colonel Lloyd's plantation
   B. Colonel Lloyd's stable or carriage-house
   C. Colonel Lloyd's mansion
   D. Colonel Lloyd's garden

5. How do **Young Barney** and **Old Barney** have particularly challenging jobs?

Finding the Irrelevant Idea

Directions: Each sentence below makes a **claim or argument** that is backed up by the four pieces of evidence that follow. **One** of those pieces of evidence is irrelevant and must be eliminated! Circle the letter that corresponds to the **irrelevant** piece of data.

1. **CLAIM:** Dogs make better pets than cats.

   A. Dogs are more loyal and friendlier than cats.
   B. Cats kill small animals, like birds and mice, and leave them around the house.
   C. A dog's bark will help keep away robbers and home invaders.
   D. Cats can occupy themselves, while dogs require more of your time and energy.

2. **CLAIM:** We should not put solar panels on the side of major highways.

   A. Solar panels are an eyesore.
   B. If someone drove off the road, they could be injured hitting one of the panels.
   C. Solar energy is not worth the time and energy it takes to harvest.
   D. Maintaining those panels so close to the road puts maintenance technicians at risk.

ACTIVITIES

3. **CLAIM:** Mr. Almond is the best 8th grade English teacher.

   A. Mr. Almond's students always do very well on standardized tests.
   B. Mr. Almond coached the boys' soccer team to a state championship last year.
   C. Mr. Almond always works to create tight bonds with his students.
   D. Mr. Almond's students get to do a ton of fun and creative projects.

4. **CLAIM:** The 8th graders shouldn't have to go on a joint field trip with the 2nd graders.

   A. Supervising two grades at the same time will be a major challenge for teachers.
   B. The two grades are unlikely to be learning similar material, so it's doubtful both groups would get a lot out of the same field trip.
   C. Watching out for the 2nd graders will be too much responsibility for the 8th graders, and they won't get to enjoy the trip.
   D. The 8th graders won the fundraising drive and were promised a field trip of their own.

FITNESS

Please be aware of your environment and be safe at all times. If you cannot do an exercise, just try your best.

Repeat these exercises **4 ROUNDS**

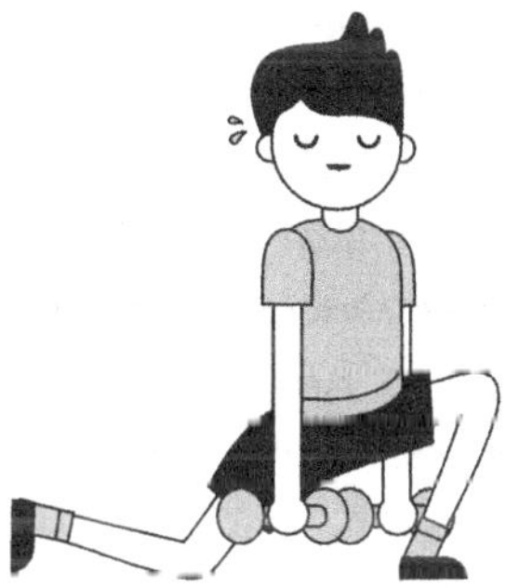

2 - Lunges: 8 times to each leg.
Note: Use your body weight or books as weight to do leg lunges.

4 - Run: 50m
Note: Run 25 meters to one side and 25 meters back to the starting position.

3 - Plank: 20 sec.

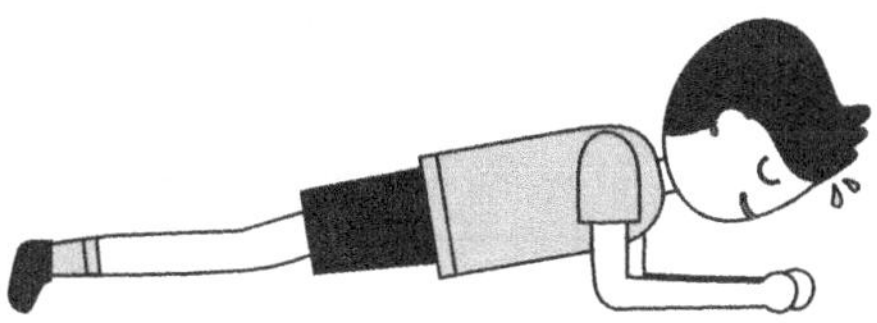

1 - Abs: 20 times

READING PASSAGE

From "Narrative of the Life of Frederick Douglass, An American Slave"

By Frederick Douglass

(Continued from Day 1's Passage)

To describe the wealth of Colonel Lloyd would be almost equal to describing the riches of Job. He kept from ten to fifteen house-servants. He was said to own a thousand slaves, and I think this estimate quite within the truth. Colonel Lloyd owned so many that he did not know them when he saw them; nor did all the slaves of the out-farms know him. It is reported of him, that, while riding along the road one day, he met a colored man, and addressed him in the usual manner of speaking to colored people on the public highways of the south: "Well, boy, whom do you belong to?" "To Colonel Lloyd," replied the slave. "Well, does the colonel treat you well?" "No, sir," was the ready reply. "What, does he work you too hard?" "Yes, sir." "Well, don't he give you enough to eat?" "Yes, sir, he gives me enough, such as it is."

The colonel, after ascertaining where the slave belonged, rode on; the man also went on about his business, not dreaming that he had been conversing with his master. He thought, said, and heard nothing more of the matter, until two or three weeks afterwards. The poor man was then informed by his overseer that, for having found fault with his master, he was now to be sold to a Georgia trader. He was immediately chained and handcuffed; and thus, without a moment's warning, he was snatched away, and forever sundered, from his family and friends, by a hand more unrelenting than death. This is the penalty of telling the truth, of telling the simple truth, in answer to a series of plain questions.

It is partly in consequence of such facts, that slaves, when inquired of as to their condition and the character of their masters, almost universally say they are contented, and that their masters are kind. The slaveholders have been known to send in spies among their slaves, to ascertain their views and feelings in regard to their condition. The frequency of this has had the effect to establish among the slaves the maxim, that a still tongue makes a wise head. They suppress the truth rather than take the consequences of telling it, and in so doing prove themselves a part of the human family. If they have any thing to say of their masters, it is generally in their masters' favor, especially when speaking to an untried man. I have been frequently asked, when a slave, if I had a kind master, and do not remember ever to have given a negative answer; nor did I, in pursuing this course, consider myself as uttering what was absolutely false; for I always measured the kindness of my master by the standard of kindness set up among slaveholders around us.

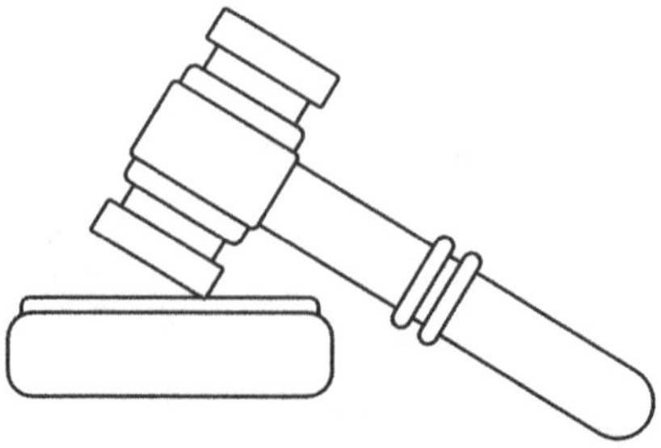

WEEK 10
DAY 2

OVERVIEW OF ENGLISH CONCEPTS

READING PASSAGE

1. Based on Paragraphs 1 & 2, **what series of events** would frequently lead to a slave being sold to a trader in Georgia?

2. Based on today's passage (and the last few from this book), how would you describe **Col. Lloyd's personality**?

3. In this passage, Douglass marvels that slaves were frequently punished for...

A. Complaining
B. Requesting different work
C. Trying to run away
D. Telling the truth

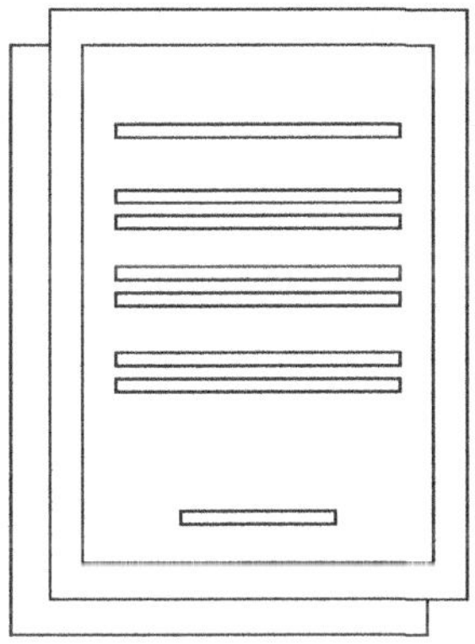

4. How could Frederick Douglass say that his master was **kind**?

A. It was true.
B. It was relatively true, compared to other slaveowners.
C. Colonel Lloyd gave Douglass special privileges.
D. Colonel Lloyd did an important favor for Douglass.

5. According to Douglass, **why** were slaves asked to describe "their condition" in a very complex and dangerous situation?

ACTIVITIES

Finding the Irrelevant Idea

Directions: Each sentence below makes a **claim or argument** that is backed up by the four pieces of evidence that follow. Each of these situations is **drawn from the text you've read.** **One** of those pieces of evidence is irrelevant and must be eliminated! Circle the letter that corresponds to the **irrelevant** piece of data.

1. **CLAIM:** Colonel Lloyd is a cruel master.

   A. Col. Lloyd tries to trick slaves into admitting he is a bad master, then sends them away.
   B. Frederick Douglass writes that Col. Lloyd was kind by the standards of other slave masters.
   C. Col. Lloyd had a fence painted with tar specifically so he could catch and punish slaves for eating fruit.
   D. Col. Lloyd yells at the slaves who care for his horses, even when they've done nothing wrong.

2. **CLAIM:** Colonel Lloyd's home plantation (the Great Farm House) is the best place to work in the area.

   A. There is less exhausting field work on the home plantation.
   B. The home plantation is a beautiful place.
   C. The slaves on the home plantation are not treated as harshly.
   D. There is a greater variety of things to do on the home plantation.

3. **CLAIM:** Slave owners like Colonel Lloyd work to prevent slaves from forming strong bonds with each other.

   A. Slaves are never allowed to work with members of their real family.
   B. Slave children are frequently separated from their mothers.
   C. Slaves can be sent away the second they are heard complaining about conditions with each other.
   D. The masters make the slaves compete and double-cross each other to get the best work.

4. **CLAIM:** Lying is an important part of daily life for slaves in the narrative.

   A. Honesty often leads to punishment or separation from others.
   B. The slaves must act as though their masters are just.
   C. Slaves must compete with each other for the best work assignments, which breeds dishonesty.
   D. The slaves must lie to each other and pretend they are happy.

FITNESS

Please be aware of your environment and be safe at all times. If you cannot do an exercise, just try your best.

Repeat these exercises **4 ROUNDS**

1 - Squats: 20 times.
Note: imagine you are trying to sit on a chair.

2 - Side Bending: 15 times to each side. Note: try to touch your feet.

3 - Tree Pose:
Stay as long as possible.
Note: do the same with the other leg.

MATH

Probability

Identify the probability of the events being independent or dependent.

1. ______________ Drawing a king from a deck of cards, then another king

2. ______________ Flipping a coin twice in a row

3. ______________ You roll a die 3 times in a row.

4. ______________ You take a marble out of a bag and then grab 3 more marbles.

5. You roll a number twice. What is the probability that you will roll a 3, then an even number?

   A. $\frac{12}{36}$ C. $\frac{6}{36}$

   B. $\frac{9}{36}$ D. $\frac{3}{36}$

6. You roll a number twice. What is the probability that you will roll an odd number, then an even number?

   A. $\frac{9}{36}$ C. $\frac{5}{36}$

   B. $\frac{7}{36}$ D. $\frac{3}{36}$

7. You roll a number twice. What is the probability that you will roll a 1, then a 1?

   A. $\frac{3}{36}$ C. $\frac{1}{36}$

   B. $\frac{2}{36}$ D. $\frac{35}{36}$

8. You roll a number twice. What is the probability that you will roll a 5, then a number less than 5?

   A. $\frac{1}{6}$ C. $\frac{1}{8}$

   B. $\frac{1}{7}$ D. $\frac{1}{9}$

You have one card for each letter. S-U-M-M-E-R-V-A-C-A-T-I-O-N

9. What is the probability you draw the letter M?

   A. $\frac{1}{7}$ C. $\frac{1}{9}$

   B. $\frac{1}{8}$ D. $\frac{1}{10}$

WEEK 10 DAY 3 MATH

10. What is the probability you draw the letter S?

A. $\frac{1}{14}$

B. $\frac{1}{12}$

C. $\frac{1}{10}$

D. $\frac{1}{8}$

11. What is the probability you draw 2 M's WITHOUT replacement?

A. $\frac{1}{14}$

B. $\frac{1}{91}$

C. $\frac{1}{10}$

D. $\frac{1}{12}$

12. What is the probability you can spell T-A-M-E by pulling the letters out in order?

A. $\frac{1}{4}$

B. $\frac{5}{14}$

C. $\frac{1}{6006}$

D. $\frac{2}{12024}$

13. What is the probability you draw the letter T?

A. $\frac{1}{14}$

B. $\frac{1}{7}$

C. $\frac{3}{14}$

D. $\frac{2}{110}$

14. What is the probability you draw the letter B?

A. $\frac{1}{14}$

B. $\frac{1}{12}$

C. 0

D. $\frac{14}{14}$

15. What is the probability you draw a vowel?

A. $\frac{1}{7}$

B. $\frac{2}{7}$

C. $\frac{3}{7}$

D. $\frac{4}{7}$

FITNESS

Please be aware of your environment and be safe at all times. If you cannot do an exercise, just try your best.

Repeat these exercises **4 ROUNDS**

1 - Bend forward: 15 times.
Note: try to touch your feet. Make sure to keep your back straight and if needed you can bend your knees.

2 - Lunges: 12 times to each leg.
Note: Use your body weight or books as weight to do leg lunges.

3 - Plank: 20 sec.

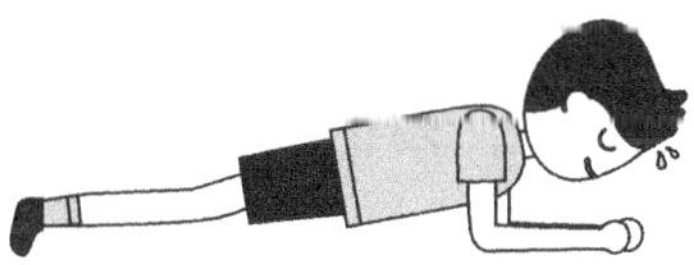

4 - Abs: 20 times

Probability

You have a bag of marbles: 2 black marbles, 3 red marbles, 10 orange marbles, 5 yellow marbles, 6 blue marbles. Find the probability that you will pick:

1. A red marble

A. $\frac{2}{26}$ B. $\frac{4}{26}$ C. $\frac{3}{26}$ D. $\frac{5}{26}$

2. An orange marble

A. $\frac{3}{26}$ B. $\frac{10}{26}$ C. $\frac{7}{36}$ D. $\frac{9}{36}$

3. Two yellow marbles WITHOUT replacement

A. $\frac{1}{36}$ B. $\frac{1}{16}$ C. $\frac{2}{36}$ D. $\frac{2}{65}$

4. A black marble then a blue marble WITHOUT replacement

A. $\frac{5}{325}$ B. $\frac{6}{325}$ C. $\frac{16}{325}$ D. $\frac{8}{325}$

5. A blue marble then a red marble WITHOUT replacement

A. $\frac{4}{325}$ B. $\frac{5}{325}$ C. $\frac{9}{325}$ D. $\frac{7}{325}$

6. A black marble

A. $\frac{6}{26}$ B. $\frac{8}{26}$ C. $\frac{10}{26}$ D. $\frac{2}{26}$

7. Two red marbles WITHOUT replacement

A. $\frac{3}{325}$ B. $\frac{2}{16}$ C. $\frac{2}{26}$ D. $\frac{2}{36}$

8. Any color but yellow

A. $\frac{20}{26}$ B. $\frac{21}{26}$ C. $\frac{22}{26}$ D. $\frac{23}{26}$

You have 15 socks: 4 blue socks, 6 black socks and 5 white socks. Find the probability you will pick:

9. A white sock

A. $\frac{1}{3}$ B. $\frac{1}{5}$ C. $\frac{1}{7}$ D. $\frac{1}{15}$

10. A blue sock

A. $\frac{1}{15}$ B. $\frac{2}{15}$ C. $\frac{3}{15}$ D. $\frac{4}{15}$

11. A black sock

A. $\frac{2}{15}$ B. $\frac{3}{15}$ C. $\frac{4}{15}$ D. $\frac{6}{15}$

WEEK 10 DAY 4 MATH

12. Two white socks WITHOUT replacement

A. $\frac{2}{21}$

B. $\frac{5}{7}$

C. $\frac{4}{15}$

D. $\frac{6}{15}$

13. Two blue socks WITHOUT replacement

A. $\frac{8}{15}$

B. $\frac{6}{15}$

C. $\frac{2}{35}$

D. $\frac{2}{15}$

14. Two black socks WITHOUT replacement

A. $\frac{1}{7}$

B. $\frac{4}{9}$

C. $\frac{5}{15}$

D. $\frac{9}{15}$

15. A white then a black sock WITHOUT replacement

A. $\frac{9}{15}$

B. $\frac{4}{15}$

C. $\frac{5}{15}$

D. $\frac{1}{7}$

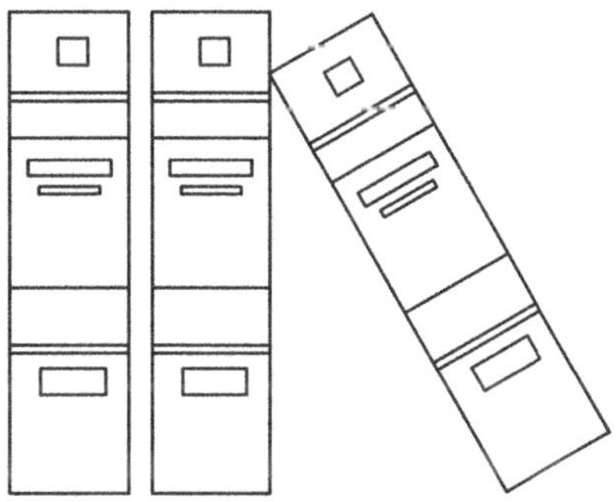

FITNESS

Please be aware of your environment and be safe at all times. If you cannot do an exercise, just try your best.

Repeat these exercises **4 ROUNDS**

1 - High Plank: 25 sec.

3 - Waist Hooping: 20 times. Note: if you do not have a hoop, pretend you have an imaginary hoop and rotate your hips 10 times.

2 - Chair: 20 sec.
Note: sit on an imaginary chair, keep your back straight.

4 - Abs: 25 times

WEEK 10 DAY 5 MATH

Probability

You have cards with the letters H-A-P-P-Y-B-I-R-T-H-D-A-Y.

1. What is the probability you draw the letter P?

A. $\frac{1}{13}$ **C.** $\frac{3}{13}$

B. $\frac{4}{13}$ **D.** $\frac{2}{13}$

2. What is the probability you draw the letter A?

A. $\frac{4}{13}$ **C.** $\frac{2}{13}$

B. $\frac{3}{13}$ **D.** $\frac{1}{13}$

3. If you draw 2 cards, how many combinations would have different letters?

A. 70 **C.** 74

B. 72 **D.** 76

4. If you draw 4 cards, how many combinations would have different letters?

A. 432 **C.** 864

B. 3024 **D.** 6048

5. If you draw 5 cards, how many combinations would have different letters?

A. 15000 **C.** 16000

B. 15120 **D.** 60480

Solve.

6. 7!

A. 4832 **C.** 5025

B. 4927 **D.** 5040

7. 4!

A. 24 **C.** 12

B. 48c **D.** 36

8. 3!

A. 4 **C.** 8

B. 6 **D.** 10

9. 6!

A. 700 **C.** 740

B. 720 **D.** 760

10. ${}_{10}C_4$

A. 200 **C.** 210

B. 205 **D.** 215

11. ${}_{14}C_2$

A. 88 **C.** 90

B. 89 **D.** 91

WEEK 10 DAY 5

MATH

12. ${}_{11}C_6$

A. 460
B. 462
C. 466
D. 486

13. ${}_{3}P_1$

A. 4
B. 3
C. 2
D. 1

14. ${}_{5}P_4$

A. 120
B. 100
C. 80
D. 60

15. ${}_{16}P_3$

A. 3060
B. 3660
C. 3630
D. 3360

YOGA

Please be aware of your environment and be safe at all times. If you cannot do an exercise, just try your best.

1 - Down Dog: 30 sec.

2 - Bend Down: 30 sec.

3 - Chair: 30 sec.

4 - Child Pose: 30 sec.

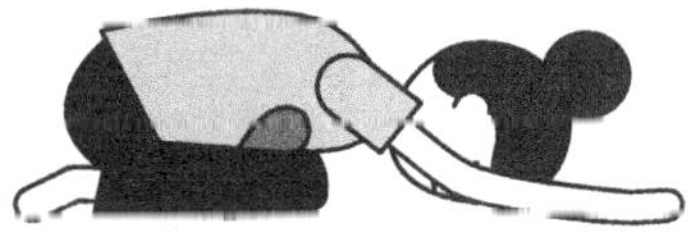

5 - Shavasana: as long as you can. Note: think of happy moments and relax your mind.

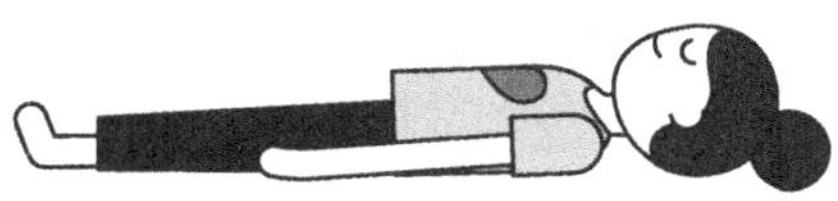

EXPERIMENT

How Bacteria Reproduce

Over the last two weeks, we looked at how cells (like the ones in your body) reproduce using ***mitosis*** *(for most cells) or* ***meiosis*** *(for gametes). Today, however, we'll be focusing on a different type of cells:* ***bacteria. Bacteria*** *are unicellular organisms, which means their entire "body" is just a single cell. However, bacteria don't have the usual organelles and organization of most other cells. For example, they don't even have a nucleus, which is considered the core of most cells!*

The word "bacteria" has a pretty bad reputation because some bacteria are responsible for the spread of ***disease*** *and* ***illness****. It's important to understand, however, that many bacteria are* ***harmless****, even* ***beneficial****, to our lives. Today, you'll create another poster to help you see how bacteria reproduce in another unique way that has nothing to do with* ***mitosis*** *or* ***meiosis****. It's known as* ***binary fission****!*

Materials:

- Your **Mitosis & Meiosis** posters from last week
- 1 large poster board
- A ruler or yardstick
- Art supplies (markers, colored pencils, etc.)
- At least two colors of yarn
- Scissors
- Glue
- An encyclopedia and/or Internet Access (for research)
- Notepaper

WEEK 10 DAY 6

EXPERIMENT

Procedure:

1. At the top of your poster board, write the title "BINARY FISSION."

2. Using your ruler and markers, divide the remainder of the poster board into **five** equal sized boxes. Label your boxes from 1-5 in this manner: **DNA Replication, Elongation of Cell, Division Begins, New Cell Walls Completed, Division Complete w/Two Daughter Cells.**

3. Using the internet (or encyclopedias) for research, look up the process of **binary fission** (you can also try "bacterial fission") and look at a variety of models to try to categorize and explain the **five steps** mentioned above. On your notepaper, write down **what's happening** in each of these stages. Try to understand both what the process "looks like" and what is happening inside the cell as it reproduces itself.

4. Once you feel like you have a general understanding of all **5** phases, use your art supplies to recreate the **5** steps of binary fission on your poster board. Use your art supplies to draw the **bacterium** that is reproducing, then cut and attach yarn to show how **genetic material** is replicated and spread.

5. Once your poster board is completed, grab your mitosis and meiosis posters from the last two weeks and compare the three side-by-side to notice the similarities and differences between the three processes.

6. Clean up your materials and answer the questions below.

WEEK 10 DAY 6 EXPERIMENT

Follow-Up Questions:

1. Based on your posters, how is **binary fission** <u>similar</u> to the processes of **mitosis** and **meiosis**?

2. Based on your posters, how is **binary fission** <u>different</u> from the processes of **mitosis** and **meiosis**?

YOGA

Please be aware of your environment and be safe at all times. If you cannot do an exercise, just try your best.

1 - Tree Pose: Stay as long as possible. Note: do on one leg then on another.

2 - Down Dog: 30 sec.

3 - Stretching: Stay as long as possible. Note: do on one leg then on another.

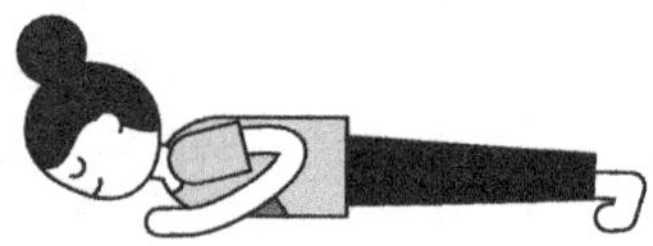

4 - Lower Plank: 20 sec. Note: Keep your back straight and body tight.

5 -Book Pose: 20 sec. Note: Keep your core tight. Legs should be across from your eyes.

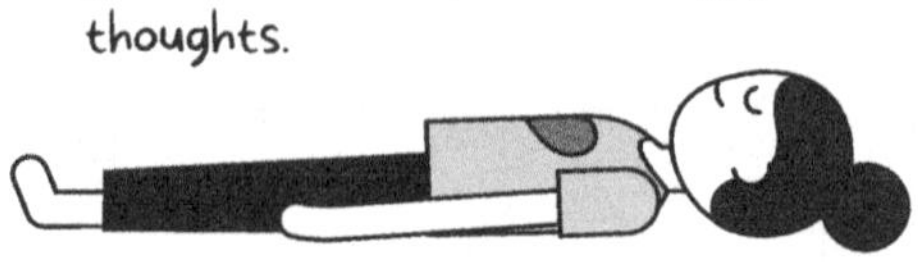

6 - Shavasana: 5 min. Note: this pose is very important and provides you with long term benefits. Try not to skip this. Close your eyes and imagine who you want to be and what your goals are! Always think happy thoughts.

LOGIC

Task:

The red car on each mini-map is moving southeast.
What is the direction of the other car on the same map?

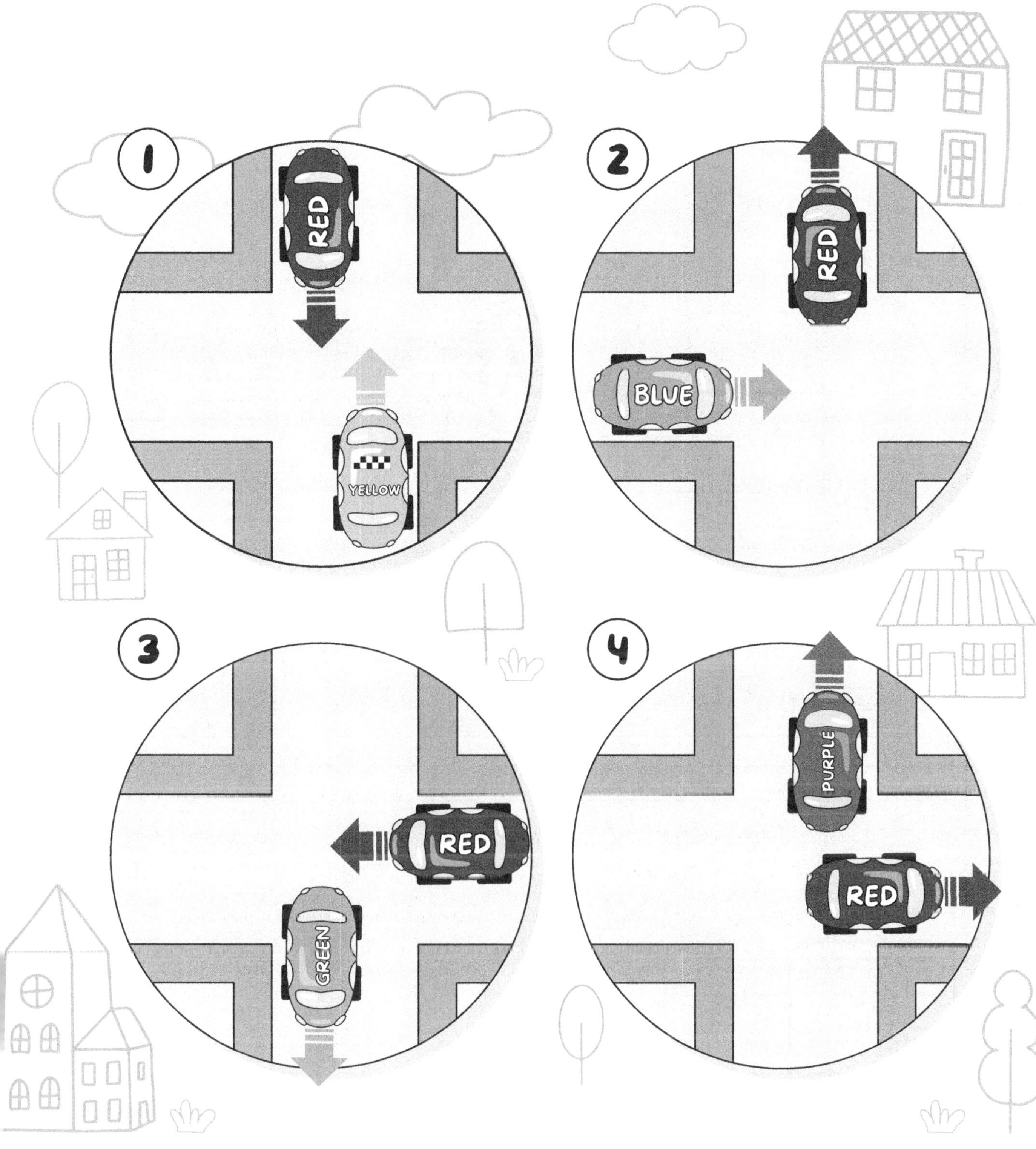

WEEK 11

ARGOPREP

WEEK 11 DAY 1

OVERVIEW OF ENGLISH CONCEPTS

Crafting Convincing Conclusions

As its title implies, the **conclusion** is the final part of an informational text. When we write, conclusions are often the thing we put the least thought and effort into. On one hand, that's natural, because conclusions usually come at the end of a long writing process, and by the time you get there, it's easy to feel like you just want to be "done." On the other hand, a truly strong conclusion is the difference between writing a great essay and writing an essay that's just fine.

Crafting a strong conclusion involves both **a clear understanding of what the conclusion needs to accomplish** (more on that in a minute) and a **strong fluency in the material that's come before it.** When you can voice both of those concerns equally in your final section, you're setting yourself up to look pretty smart and convincing to your reader.

The Three Goals of a Conclusion:

A strong conclusion, whether it's a single sentence, a designated paragraph, or a whole chapter of a longer book, should have three main goals:

- **Restate your most important main ideas** - One of the important jobs your conclusion has is to summarize what came before it. You need to remind the reader of your main idea and what other important concepts are connected to it.
- **Remind the reader why they should care** - To have a big impact on a reader, a text must seem important, relevant, and connected to their lives in some way. Your conclusion is your last chance to help your reader understand why your topic and ideas are so important.
- **End with a bang!** - Memorable texts have memorable endings - just think about your favorite movie or novel. Your conclusion should wrap up with a "**final thought**" that will keep your reader thinking about the material you covered even after they put down what you've written. Your conclusion is the last chance to put that "ear worm" into your reader!

OVERVIEW OF ENGLISH CONCEPTS

Test Driving Your Conclusion

If you want to be sure you're writing strong conclusions, here's a little activity that you can do with a partner, teacher, or family member to help you test the strength of your conclusion:

1. Create a **rough draft** of your conclusion statement. Try to be sure it **does the three jobs listed above.**
2. Print out your conclusion **by itself** (without the rest of the essay) and give it to your partner.
3. Have your partner **read the conclusion** and see if they can answer the following questions:
    a. What is this essay or paper's main topic?
    b. What are a few of the most important main ideas in the essay?
    c. How does the author want you to think or feel when you're done with the essay?
4. Once your partner's done, review their responses. If they were able to answer all three questions **correctly**, then your conclusion is probably pretty strong! If there was any confusion as to the answers of any of the three questions, you should edit and revise your conclusion to clarify those ideas and create the best possible reading experience!

READING PASSAGE

From "Narrative of the Life of Frederick Douglass, An American Slave"

By Frederick Douglass

*(**NOTE:** This passage continues from last week's text.)*

Mr. Hopkins remained but a short time in the office of overseer. Why his career was so short, I do not know, but suppose he lacked the necessary severity to suit Colonel Lloyd. Mr. Hopkins was succeeded by Mr. Austin Gore, a man possessing, in an eminent degree, all those traits of character indispensable to what is called a first-rate overseer. Mr. Gore had served Colonel Lloyd, in the capacity of overseer, upon one of the out-farms, and had shown himself worthy of the high station of overseer upon the home or Great House Farm.

Mr. Gore was proud, ambitious, and persevering. He was artful, cruel, and obdurate. He was just the man for such a place, and it was just the place for such a man. It afforded scope for the full exercise of all his powers, and he seemed to be perfectly at home in it. He was one of those who could torture the slightest look, word, or gesture, on the part of the slave, into impudence, and would treat it accordingly. There must be no answering back to him; no explanation was allowed a slave, showing himself to have been wrongfully accused. Mr. Gore acted fully up to the maxim laid down by slaveholders, — "It is better that a dozen slaves should suffer under the lash, than that the overseer should be convicted, in the presence of the slaves, of having been at fault." No matter how innocent a slave might be — it availed him nothing, when accused by Mr. Gore of any misdemeanor. To be accused was to be convicted, and to be convicted was to be punished; the one always following the other with immutable certainty. To escape punishment was to escape accusation; and few slaves had the fortune to do either, under the overseership of Mr. Gore. He was just proud enough to demand the most debasing homage of the slave, and quite servile enough to crouch, himself, at the feet of the master. He was ambitious enough to be contented with nothing short of the highest rank of overseers, and persevering enough to reach the height of his ambition. He was cruel enough to inflict the severest punishment, artful enough to descend to the lowest trickery, and obdurate enough to be insensible to the voice of a reproving conscience. He was, of all the overseers, the most dreaded by the slaves. His presence was painful, his eye flashed confusion, and seldom was his sharp, shrill voice heard, without producing horror and trembling in their ranks.

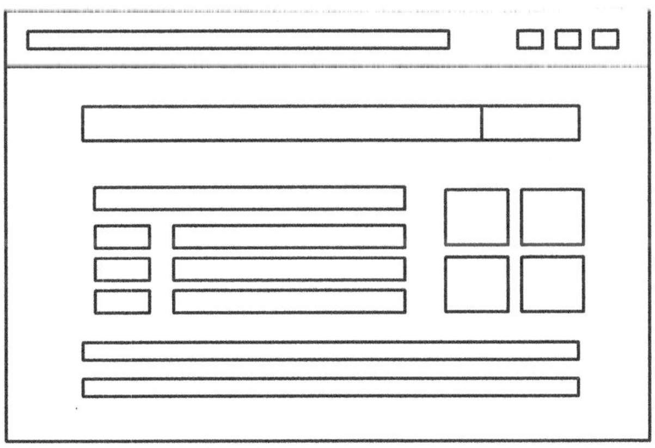

OVERVIEW OF ENGLISH CONCEPTS
READING PASSAGE

1. According to the text, why did Mr. Hopkins **probably** lose his job as **overseer**?

2. Describe **Mr. Gore's** personality:

3. How does **Mr. Gore** try to portray himself to the slaves?

   A. He wants them to think he is just and kind.
   B. He wants them to think that he is tough, but fair.
   C. He wants them to think he is absolutely correct in every situation.
   D. He wants them to think of him as a middle man between them and Col. Lloyd.

4. What **practice of Mr. Gore's** does Douglass specifically mention as unsavory?

   A. Mr. Gore frequently lies to the slaves.
   B. Mr. Gore frequently lies to Colonel Lloyd.
   C. Mr. Gore makes the slaves bow or crouch before him.
   D. Mr. Gore makes the slaves do twice as much work as the other overseers.

5. Based on all the things you've read from *The Narrative of Frederick Douglass* so far, why do you think Mr. Gore was the **most dreaded** of all the overseers?

Crafting a Conclusion. (Pt. 1)

Directions: Read the article below and underline main or important ideas as you go. On Day 2, you'll be **writing a conclusion paragraph** for this text, so your goal should be to gain a strong understanding of **what the text says**, what the **main ideas are**, and **how the author is approaching the topic** and reader.

"How to Grow Turnips"

From The Household Cyclopedia of General Information

The benefits derived from turnip husbandry are of great magnitude; light soils are cultivated with profit and facility; abundance of food is provided for man and beast; the earth is turned to the uses for which it is physically calculated, and by being suitably cleaned with this preparatory crop, a bed is provided for grass seeds, wherein they flourish and prosper with greater vigor than after any other preparation.

The first ploughing is given immediately after harvest, or as soon as the wheat seed is finished, either in length or across the field, as circumstances may seem to require. In this state the ground remains till the oat seed is finished, when a second ploughing is given to it, usually in a contrary direction to the first. It is then repeatedly harrowed, often rolled between the harrowings and every particle of root-weeds carefully picked off with the hand; a third ploughing is then bestowed, and the other operations are repeated. In this stage, if the ground has not been very foul, the seed process generally commences, but often a fourth ploughing, sometimes a fifth is necessary before the ground is sufficiently clean. Less labor, however, is necessary now than in former times, when a more regular mode of cropping was commonly followed.

The next part of the process is the sowing of the seed; this may be performed by drilling machines of different sizes and constructions, through all acting on the same principle. A machine drawn by a horse in a pair of shafts, sows two drills at a time and answers extremely well, where the ground is flat, and the drills properly made up. The weight of the machine insures a regularity of sowing hardly to be gained by those of a different size and construction. From two to three pounds of seed are sown upon the acre, though the smallest of these quantities will give many more plants in ordinary seasons than are necessary; but as the seed is not an expensive article the greater part of farmers incline to sow thick, which both provides against the danger of part of the seed perishing, and gives the young plants an advantage at the outset.

Turnips are sown from the beginning to the end of June, but the second and third weeks of the month are, by judicious farmers, accounted the most proper time. Some people have sown as early as May, and with advantage, but these early fields are apt to run to seed before winter, especially if the autumn be favorable to vegetation. As a general rule it may be laid down that the earliest sowings should be on the latest soils; plants on such soils are often long before they make any great progress, and, in the end, may be far behind those in other situations, which were much later sown. The turnip plant, indeed, does not thrive rapidly till its roots reach the dung, and the previous nourishment afforded them is often so scanty as to stunt them altogether before they get so far.

WEEK 11 DAY 1

OVERVIEW OF ENGLISH CONCEPTS
ACTIVITIES

The first thing to be done in this process is to run a horse-hoe, called a scraper, along the intervals, keeping at such a distance from the young plants that they shall not be injured; this operation destroys all the annual weeds which have sprung up, and leaves the plants standing in regular stripes or rows. The hand hoeing then commences, by which the turnips are all singled out at a distance of from eight to twelve inches, and the redundant ones drawn into the spaces between the rows. The singling out of the young plants is an operation of great importance, for an error committed in this process can hardly be afterwards rectified. Boys and girls are always employed as hoers; but a steady and trusty man-servant is usually set over them to see that the work is properly executed.

In eight or ten days, or such a length of time as circumstances may require, a horse-hoe of a different construction from the scraper is used. This, in fact, is generally a small plough, of the same kind with that commonly wrought, but of smaller dimensions. By this implement, the earth is pared away from the sides of the drills, and a sort of new ridge formed in the middle of the former interval. The hand-hoers are again set to work, and every weed and superfluous turnip is cut up; afterwards the horse-hoe is employed to separate the earth, which it formerly threw into the furrows, and lay it back to the sides of the drills. On dry lands this is done by the scraper, but where the least tendency to moisture prevails, the small plough is used, in order that the furrows may be perfectly cleaned out. This latter mode, indeed, is very generally practiced.

FITNESS

Please be aware of your environment and be safe at all times. If you cannot do an exercise, just try your best.

Repeat these exercises **4 ROUNDS**

1 - Abs: 20 times

2 - Lunges: 8 times to each leg.
Note: Use your body weight or books as weight to do leg lunges.

3 - Plank: 20 sec.

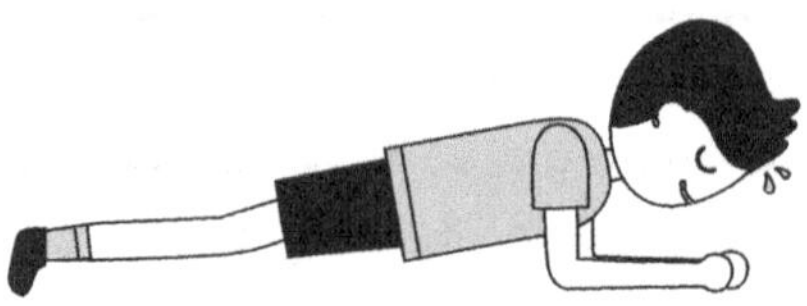

4 - Run: 50m
Note: Run **25** meters to one side and **25** meters back to the starting position.

From "Narrative of the Life of Frederick Douglass, An American Slave"

By Frederick Douglass

(Continued from Day 1's Passage)
As to my own treatment while I lived on Colonel Lloyd's plantation, it was very similar to that of the other slave children. I was not old enough to work in the field, and there being little else than field work to do, I had a great deal of leisure time. The most I had to do was to drive up the cows at evening, keep the fowls out of the garden, keep the front yard clean, and run of errands for my old master's daughter, Mrs. Lucretia Auld. The most of my leisure time I spent in helping Master Daniel Lloyd in finding his birds, after he had shot them. My connection with Master Daniel was of some advantage to me. He became quite attached to me, and was a sort of protector of me. He would not allow the older boys to impose upon me, and would divide his cakes with me.

I was seldom whipped by my old master, and suffered little from anything else than hunger and cold. I suffered much from hunger, but much more from cold. In hottest summer and coldest winter, I was kept almost naked — no shoes, no stockings, no jacket, no trousers, nothing on but a coarse tow linen shirt, reaching only to my knees. I had no bed. I must have perished with cold, but that, the coldest nights, I used to steal a bag which was used for carrying corn to the mill. I would crawl into this bag, and there sleep on the cold, damp, clay floor, with my head in and feet out. My feet have been so cracked with the frost, that the pen with which I am writing might be laid in the gashes.

We were not regularly allowanced. Our food was coarse corn meal boiled. This was called mush. It was put into a large wooden tray or trough, and set down upon the ground. The children were then called, like so many pigs, and like so many pigs they would come and devour the mush; some with oyster-shells, others with pieces of shingle, some with naked hands, and none with spoons. He that ate fastest got most; he that was strongest secured the best place; and few left the trough satisfied.

I was probably between seven and eight years old when I left Colonel Lloyd's plantation. I left it with joy. I shall never forget the ecstasy with which I received the intelligence that my old master (Anthony) had determined to let me go to Baltimore, to live with Mr. Hugh Auld, brother to my old master's son-in-law, Captain Thomas Auld. I received this information about three days before my departure. They were three of the happiest days I ever enjoyed. I spent the most part of all these three days in the creek, washing off the plantation scurf, and preparing myself for my departure.

READING PASSAGE

1. What were some of **Frederick's** responsibilities on the Lloyd plantation?

2. Describe the relationship between **Frederick** and **Master Daniel Lloyd**:

3. According to Frederick, what was the worst personal hardship for him during his time on Col. Lloyd's plantation?

   A. The beatings
   B. Hunger
   C. Being separated from his mother
   D. Cold

4. What did Frederick use a corn sack as?

   A. A shirt
   B. A sleeping bag
   C. A backpack
   D. A secret hiding place

5. Why were Frederick's last three days on Col. Lloyd's plantation so **happy**?

Crafting a Conclusion

Directions: Review the text about turnips that you read on Day 1 and **double-check your underlined information** to make sure you understand the **main ideas** of the passage. Once you feel confident that you have a strong grasp on the passage, use the lines below to write a conclusion paragraph that could go at the end of that article. Remember to make sure your conclusion paragraph...

- **Restates the most important main ideas**
- **Reminds the reader why they should care**
- **Ends things with a bang!**

WEEK 11 DAY 2

OVERVIEW OF ENGLISH CONCEPTS ACTIVITIES

FITNESS

Please be aware of your environment and be safe at all times. If you cannot do an exercise, just try your best.

Repeat these exercises **4 ROUNDS**

1 - Squats: 20 times.
Note: imagine you are trying to sit on a chair.

2 - Side Bending: 15 times to each side. Note: try to touch your feet.

3 - Tree Pose: Stay as long as possible.
Note: do the same with the other leg.

MATH

Review

Use <, > or = to complete each statement.

1. $\frac{7}{8}$ ___ $\frac{7}{9}$

2. $\frac{2}{3}$ ___ $\frac{20}{46}$

3. $\frac{1}{8}$ ___ $\frac{7}{56}$

4. Which number is the smallest?

A. 0.5
B. $\frac{1}{10}$
C. 0
D. $-\frac{1}{4}$

5. Which number is the largest?

A. $\frac{1}{6}$
B. -0.6
C. $\frac{3}{5}$
D. $\frac{2}{3}$

6. What is $\frac{3}{8}$ as a decimal?

A. 0.375
B. 1.375
C. 3.75
D. 0.75

7. $3\frac{3}{4} \times \left(2\frac{4}{5}\right)$

A. $\frac{42}{4}$
B. $\frac{21}{4}$
C. $\frac{21}{2}$
D. $\frac{42}{2}$

8. $\frac{4}{m} \div \frac{5m}{9}$

A. $\frac{20}{9}$
B. $\frac{36}{5m^2}$
C. $36m^2$
D. $\frac{9}{m}$

9. Simplify $-(2x^2y)^3$

A. $8x^6y^3$
B. $-8x^6y^3$
C. $8x^6$
D. $24y^3$

10. $-6b = 3\frac{1}{3}$

A. $b = \frac{9}{2}$
B. $b = -\frac{9}{2}$
C. $b = \frac{2}{9}$
D. $b = -\frac{5}{9}$

11. What is $0.\overline{7}$ as a fraction?

A. $\frac{7}{9}$
B. $\frac{2}{7}$
C. $\frac{1}{7}$
D. $\frac{4}{9}$

WEEK 11 DAY 3

MATH

Review

12. $\frac{2}{3} - (-14) =$

A. $13\frac{2}{3}$

B. $-14\frac{2}{3}$

C. $14\frac{2}{3}$

D. $-13\frac{2}{3}$

13. What is the GCF of 16 and 20?

A. 2
B. 4
C. 12
D. 10

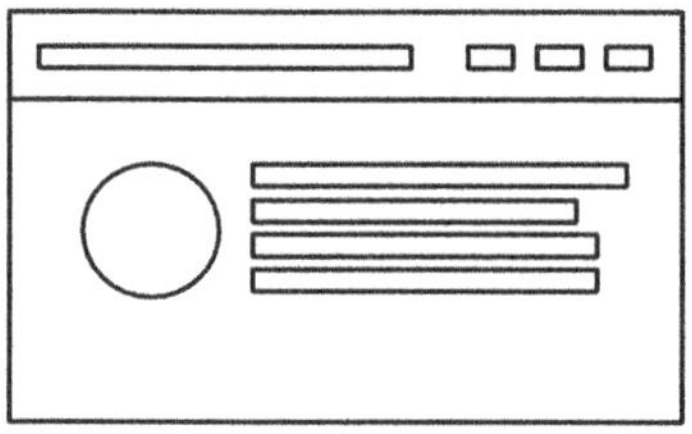

14. $\frac{3}{8}y = -15$

A. $y = -40$
B. $y = 40$
C. $y = -20$
D. $y = 20$

15. $2 \times |15 - 22| + 4^2$

A. 20
B. 25
C. 27
D. 30

FITNESS

Please be aware of your environment and be safe at all times. If you cannot do an exercise, just try your best.

Repeat these exercises **4 ROUNDS**

2 - Lunges: 12 times to each leg.
Note: Use your body weight or books as weight to do leg lunges.

3 - Plank: 20 sec.

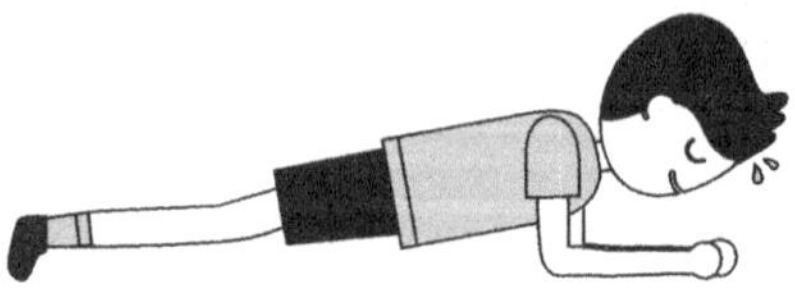

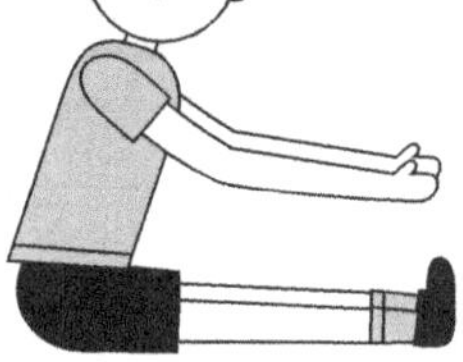

1 - Bend forward: 15 times.
Note: try to touch your feet. Make sure to keep your back straight and if needed you can bend your knees.

4 - Abs: 20 times

WEEK 11 DAY 4 MATH

Review

1. What is the unit rate if a car travels 96 miles on 4 tanks of gas?

   A. 20
   B. 22
   C. 24
   D. 26

2. What is the unit rate if a bicyclist travels 33 miles in 5 hours?

   A. 6.2
   B. 6.4
   C. 6.6
   D. 6.8

3. Three cans of cat food sell for $1.05. What is the price of one can?

   A. 33 cents
   B. 35 cents
   C. 37 cents
   D. 39 cents

4. A photo that measures 5 x 7 is enlarged to 15 x 21. How much larger is the second photo?

   A. 3 times
   B. 4 times
   C. 5 times
   D. 7 times

5. A student reads 60 pages in 2.5 hours. How many pages does he read in another 4 hours?

   A. 96 pages
   B. 192 pages
   C. 32 pages
   D. 16 pages

6. What percent of 400 is 20?

   A. 1%
   B. 2%
   C. 4%
   D. 5%

7. What is 45% of 12?

   A. 5.1
   B. 5.3
   C. 5.2
   D. 5.4

8. What is 20% of 24?

   A. 4.2
   B. 4.4
   C. 4.8
   D. 4.6

9. What percent of 75 is 15?

   A. 10%
   B. 20%
   C. 30%
   D. 40%

10. 60% percent of what number is 120?

   A. 200
   B. 220
   C. 240
   D. 260

11. If a shirt is reduced to $36 from $60, what is the percent decrease?

   A. 11%
   B. 22%
   C. 33%
   D. 40%

WEEK 11 DAY 4 MATH

Review

12. If a shirt is marked up from **$18** to **$24**, what is the percent increase?

A. 11%
B. 22%
C. 33%
D. 66%

13. A salesman makes a **15%** commission and sells **$2400** worth of merchandise. How much is his commission?

A. 300
B. 360
C. 420
D. 480

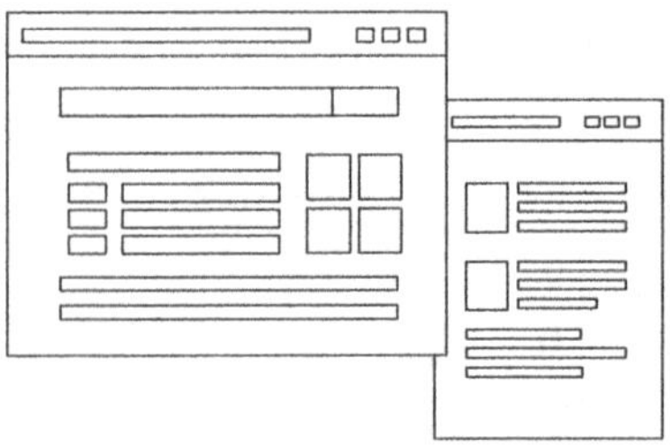

14. Elise buys a new purse for **$224**. If she has to pay **6%** tax, what is the total cost?

A. $237.44
B. $247.88
C. $257.66
D. $268.22

15. Sean and three friends go out to lunch. Their total bill is **$34**. If they want to give a **15%** tip and each want to pay evenly, how much do each of them have to pay?

A. $9.78
B. $9.79
C. $9.80
D. $9.81

FITNESS

Please be aware of your environment and be safe at all times. If you cannot do an exercise, just try your best.

Repeat these exercises **4 ROUNDS**

1 - High Plank: 25 sec.

3 - Waist Hooping: 20 times. Note: if you do not have a hoop, pretend you have an imaginary hoop and rotate your hips 10 times.

2 - Chair: 20 sec.
Note: sit on an imaginary chair, keep your back straight.

4 - Abs: 25 times

WEEK 11 DAY 5 MATH

Review

1. $33 - 4x = 13$

   A. $x = 2$
   B. $x = 3$
   C. $x = 4$
   D. $x = 5$

2. $5x + 7 = 47$

   A. $x = 7$
   B. $x = 8$
   C. $x = 9$
   D. $x = 10$

3. $4b - b = 36$

   A. $b = 12$
   B. $b = 13$
   C. $b = 14$
   D. $b = 15$

4. $2a - 24 = -8$

   A. $a = 6$
   B. $a = 7$
   C. $a = 8$
   D. $a = 9$

5. $10 - 5x = -115$

   A. $x = 10$
   B. $x = 15$
   C. $x = 20$
   D. $x = 25$

6. $8x - 11 = 37$

   A. $x = 4$
   B. $x = 6$
   C. $x = 8$
   D. $x = 10$

7. $14x - 11 = 3$

   A. $x = 1$
   B. $x = 2$
   C. $x = 3$
   D. $x = 4$

8. $3a - 5a = -28$

   A. $a = 12$
   B. $a = 14$
   C. $a = 16$
   D. $a = 18$

9. $14 + 7c = 147$

   A. $c = 13$
   B. $c = 15$
   C. $c = 17$
   D. $c = 19$

10. $21 - 5d = -34$

   A. $d = 7$
   B. $d = 9$
   C. $d = 11$
   D. $d = 13$

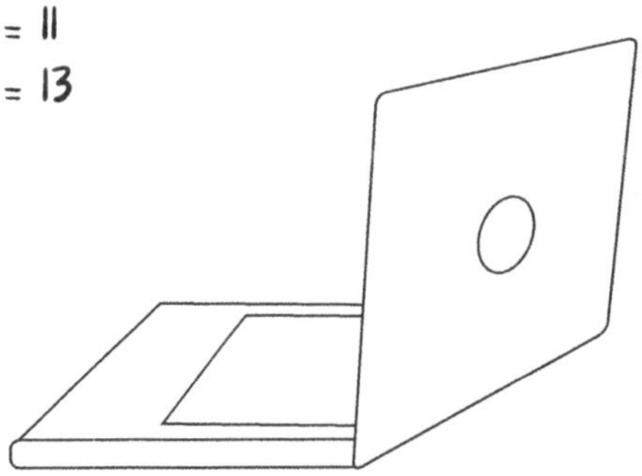

WEEK 11 DAY 5

MATH

Review

11. $2a - 3 > 1$

A. $a > 1$
B. $a > 2$
C. $a > 3$
D. $a > 4$

12. $-6f + 12 < 6$

A. $f > 1$
B. $f > 2$
C. $f > 3$
D. $f > 4$

13. $\frac{4}{9}y + 11 < 16\frac{1}{3}$

A. $y < 12$
B. $y < 14$
C. $y < 16$
D. $y < 18$

14. $12k - 8 > 136$

A. $k > 10$
B. $k > 12$
C. $k > 14$
D. $k > 16$

15. $x + 5x + 13 < 67$

A. $x < 7$
B. $x < 8$
C. $x < 9$
D. $x < 10$

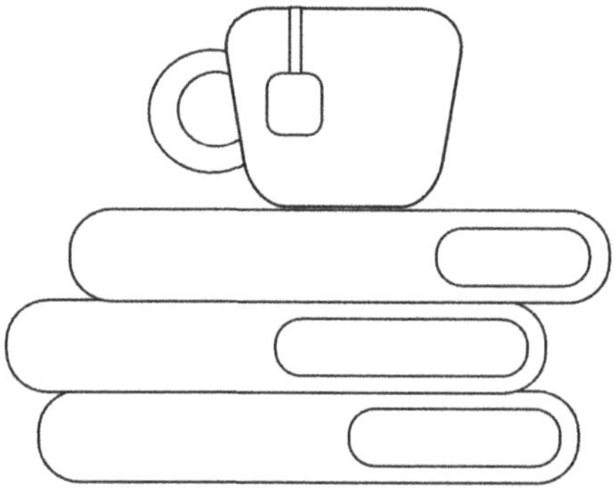

YOGA

Please be aware of your environment and be safe at all times. If you cannot do an exercise, just try your best.

1 - Down Dog: 30 sec.

2 - Bend Down: 30 sec.

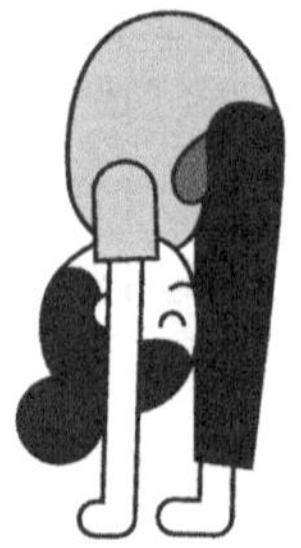

3 - Chair: 30 sec.

4 - Child Pose: 30 sec.

5 - Shavasana: as long as you can. Note: think of happy moments and relax your mind.

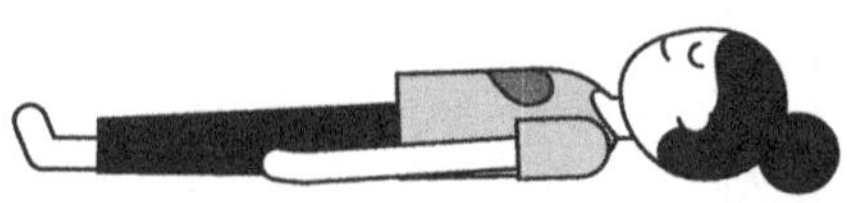

WEEK 11 DAY 6

EXPERIMENT

How Viruses Reproduce

*Over the last three weeks, you've looked at how different **cells** and **microorganisms** reproduce through various processes like **mitosis**, **meiosis**, and **bacterial fission**. This week, we'll take one final look into the world of microbiology, studying one final microbe: a **virus**. Viruses are dangerous because they can't reproduce on their own - they can only create new viruses by infecting a **host**. Once a virus is inside an organism's body, it basically steals that organism's cells and uses them as incubators to reproduce. That process is very destructive to cells, which is why viruses can make people (and animals) so incredibly sick!*

*This week, we'll create one more poster to look at a process called **viral replication**. Now that you have a greater understanding of cells and cell reproduction, you'll be able to appreciate how destructive and dangerous viruses can be! As you work on this activity, try to think both about how the virus replicates but also how that process affects the host cell.*

Materials:

- Your **Mitosis, Meiosis, and Binary Fission** posters from last week
- 1 large poster board
- A ruler or yardstick
- Art supplies (markers, colored pencils, etc.)
- At least two colors of yarn
- Scissors
- Glue
- An encyclopedia and/or Internet Access (for research)
- Notepaper

EXPERIMENT

Procedure:

1. At the top of your poster board, write the title "VIRAL REPLICATION"

2. Using your ruler and markers, divide the remainder of the poster board into **five** equal sized boxes. Label your boxes from 1-5 in this manner: **Attachment, Entry, Replication, Assembly,** and **Release.**

3. Using the internet (or encyclopedias) for research, look up the process of **viral replication** (you can also try "viral reproductions") and look at a variety of models to try to categorize and explain the **five steps** mentioned above. On your notepaper, write down **what's happening** in each of these stages. Try to understand both what the process "looks like" and what is happening inside the cell as the virus replications.

4. Once you feel like you have a general understanding of all **5** phases, use your art supplies to recreate the **5** steps of viral replication on your poster board. Use your art supplies to draw the **bacterium** that is the "host cell," then cut and attach yarn to show how **viral genetic material** invades and repurposes the cell.

5. Once your poster board is complete, grab your mitosis, meiosis, and binary fission posters from the last three weeks and compare the four side-by-side to notice the similarities and differences between the four processes.

6. Clean up your materials and answer the questions below.

WEEK 11 DAY 6 EXPERIMENT

Follow-Up Questions:

1. How are viruses **different** from the other cells you've looked at in recent activities?

2. Based on what you saw as you created your posters, why are **viruses** potentially very dangerous and destructive?

YOGA

Please be aware of your environment and be safe at all times. If you cannot do an exercise, just try your best.

2 - Down Dog: 30 sec.

3 - Stretching: Stay as long as possible.
Note: do on one leg then on another.

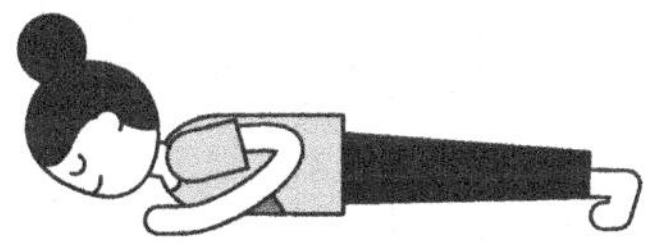

4 - Lower Plank: 20 sec.
Note: Keep your back straight and body tight.

1 - Tree Pose: Stay as long as possible.
Note: do on one leg then on another.

6 - Shavasana: 5 min.
Note: this pose is very important and provides you with long term benefits. Try not to skip this. Close your eyes and imagine who you want to be and what your goals are! Always think happy thoughts.

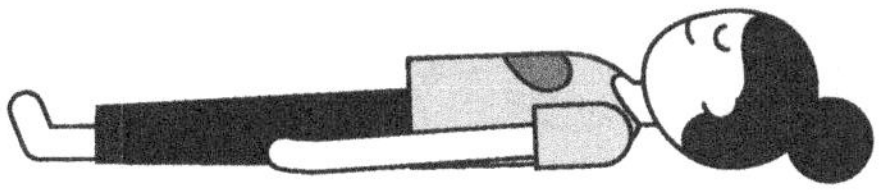

5 -Book Pose: 20 sec.
Note: Keep your core tight. Legs should be across from your eyes.

WEEK 11 DAY 7

MAZE

Task: Help the taxi to find the way to the house.

WEEK 12

ARGOPREP

OVERVIEW OF ENGLISH CONCEPTS

Evaluating the Quality of Texts

As we wrap up our summer studies, it's time to think about one of the most important ideas when it comes to reading and writing: is the text actually any good? Thinking about the **quality** of texts can be tricky, especially when you're young. As you get closer to high school, however, you will need to start **evaluating** texts, not just reading them.

Evaluating a text means studying it with an eye toward not just its meaning, but also its **worth, value, and quality**. This can be complex at first because it requires **thinking beyond yourself** and considering how good a job the text does at accomplishing its original goal. Your own personal opinions are crucial to informing this process, but they are **not** the be-all-end-all. Let's take a look at some qualities you can study or assess to help you **evaluate the quality** of texts:

What Makes a Text "Good?"

Clarity: Good texts are **clear** for the reader and **easy to follow**; Bad texts are confusing or unclear.

Engagement: Good texts are interesting and **hold your attention**; Bad texts are boring and difficult to get through.

Takeaways: Good texts communicate ideas or information that you can **apply or connect to your own life**; Bad texts leave it unclear how information is relevant and how you can put it into practice.

Activates Intellectual Curiosity: Good texts raise new questions and **make you want to explore and learn more**; Bad texts make you tired of a topic and never want to learn more about it again.

A Good Example of its Genre: Good texts provide fun and exciting **spins on genre tropes** to create something new and unique; Bad texts rely heavily on genre topes without the imagination or creativity needed to make them interesting.

Memorableness: Good texts are memorable and make you want to read more on that topic or by that author; Bad texts are forgettable or only easy to remember as "that terrible book I had to read!"

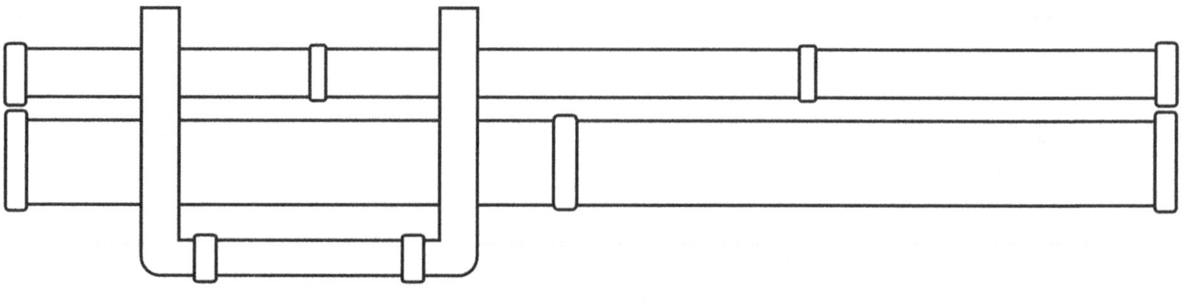

WEEK 12 DAY 1

OVERVIEW OF ENGLISH CONCEPTS

Tips for Evaluating Texts:

- Evaluate **as you read**; don't wait for the end!
    - If you wait until you're **done** with a text to start figuring out if it's good or valuable, you've gone too far! Constantly keep track in your head or (ideally) your notes of how a text is developing.
        - Write down each thing that feels **strong**, **interesting**, **or important to you**
        - Also, keep notes of things that seem **dumb**, **unimportant**, or **fail to make sense**
- Try to remain **objective** as you evaluate
    - Don't just judge the book based on what you **like or don't like**, try to think **beyond yourself** and consider the text **as a text**
    - Try to think about the **six standards we listed above**
        - Those qualities will keep you **grounded** and as close to **objective** as possible
- If you think a text is really **bad**, find ways to have fun with it!
    - As much as it stinks, you can't stop reading assigned text just because you don't like it.
    - Keep track of all the things that are "bad," "dumb," or poorly explained. Later on, you can share those with others to illustrate what a bad text it was.

From "Narrative of the Life of Frederick Douglass, An American Slave"

By Frederick Douglass

I was probably between seven and eight years old when I left Colonel Lloyd's plantation. I left it with joy. I shall never forget the ecstasy with which I received the intelligence that my old master (Anthony) had determined to let me go to Baltimore, to live with Mr. Hugh Auld, brother to my old master's son-in-law, Captain Thomas Auld. I received this information about three days before my departure. They were three of the happiest days I ever enjoyed. I spent the most part of all these three days in the creek, washing off the plantation scurf, and preparing myself for my departure.

The pride of appearance which this would indicate was not my own. I spent the time in washing, not so much because I wished to, but because Mrs. Lucretia had told me I must get all the dead skin off my feet and knees before I could go to Baltimore; for the people in Baltimore were very cleanly, and would laugh at me if I looked dirty. Besides, she was going to give me a pair of trousers, which I should not put on unless I got all the dirt off me. The thought of owning a pair of trousers was great indeed! It was almost a sufficient motive, not only to make me take off what would be called by pig-drovers the mange, but the skin itself. I went at it in good earnest, working for the first time with the hope of reward.

The ties that ordinarily bind children to their homes were all suspended in my case. I found no severe trial in my departure. My home was charmless; it was not home to me; on parting from it, I could not feel that I was leaving anything which I could have enjoyed by staying. My mother was dead, my grandmother lived far off, so that I seldom saw her. I had two sisters and one brother, that lived in the same house with me; but the early separation of us from our mother had well nigh blotted the fact of our relationship from our memories. I looked for home elsewhere, and was confident of finding none which I should relish less than the one which I was leaving. If, however, I found in my new home hardship, hunger, whipping, and nakedness, I had the consolation that I should not have escaped any one of them by staying. Having already had more than a taste of them in the house of my old master, and having endured them there, I very naturally inferred my ability to endure them elsewhere, and especially at Baltimore; for I had something of the feeling about Baltimore that is expressed in the proverb, that "being hanged in England is preferable to dying a natural death in Ireland." I had the strongest desire to see Baltimore.

OVERVIEW OF ENGLISH CONCEPTS

READING PASSAGE

1. Do you think **Lucretia** was looking out for Frederick in this passage? **Why** or why not?

2. Why did Frederick **clean himself** so aggressively before leaving the plantation?

3. What does Frederick mean when he says, "**The ties that ordinarily bind** children to their homes were all suspended in my case"?

   A. He has never been tied up with rope or chains.
   B. He has no family and no love of his home.
   C. He received a suspended sentence in a court trial.
   D. Nobody ever worried or cared about him.

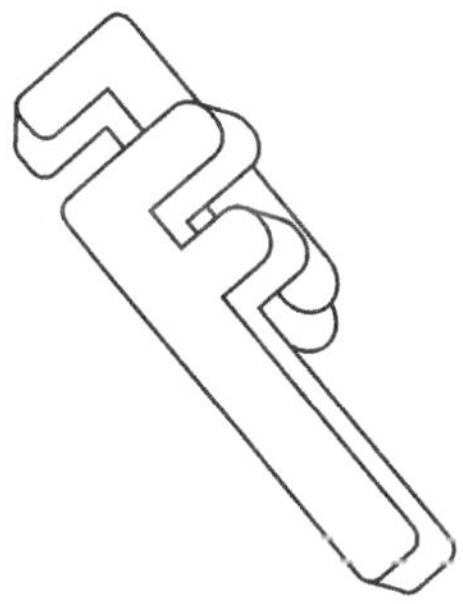

4. Why is Frederick **not** close to his siblings?

   A. They were all separated from their mother at an early age.
   B. They enjoy working on the plantation.
   C. They are fearful of Mr. Gore and Colonel Lloyd.
   D. They believe he is too close with Daniel Lloyd and Ms. Lucretia.

5. Why does Frederick want to see **Baltimore** so badly?

Evaluating a Strong Text

Directions: Think about a text that is **one of your favorites** and that you consider to be **a very strong text**. It could be a book, a short story, a movie, an episode of a TV show, etc. You could use an **informational text** or a **literary text** for this activity! Write the title of that text on the line below, then answer the following questions to help you think and analyze what makes that text so **good**!

Text Title: ______________________________

- Just generally, **what do you enjoy** about this text?

  ____

  ____

- Is the text **clear** throughout? What does the author do to make sure you **understand what is going on** throughout the book?

  ____

  ____

  ____

  ____

- How does the author **hold your interest** throughout the book? What tricks are used to **engage** you?

  ____

  ____

  ____

  ____

OVERVIEW OF ENGLISH CONCEPTS

ACTIVITIES

- What **takeaways** does the author include in this text? How does it feel connected or applicable to your life? What can you **learn** from this text?

- How does this author **access or activate your intellectual curiosity**? What do they make you want to learn or know more about?

OVERVIEW OF ENGLISH CONCEPTS
ACTIVITIES

- **Why** is this text a **good example of its genre**? What **innovative** or **interesting** techniques does it use?

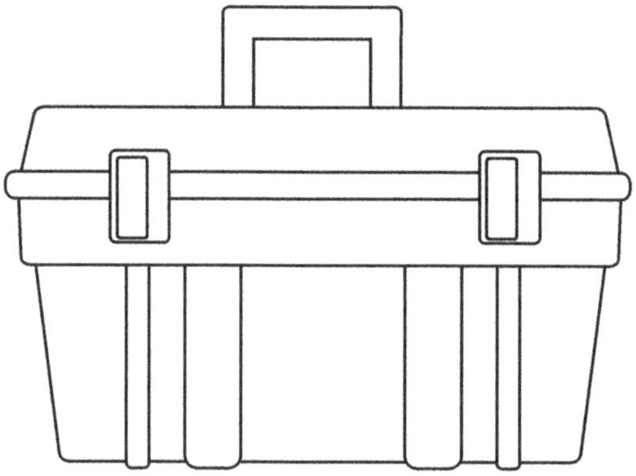

FITNESS

Please be aware of your environment and be safe at all times. If you cannot do an exercise, just try your best.

Repeat these exercises **4 ROUNDS**

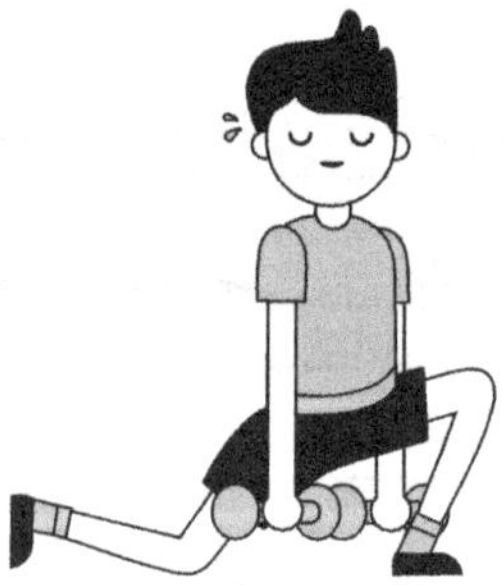

2 - Lunges: 8 times to each leg.
Note: Use your body weight or books as weight to do leg lunges.

4 - Run: 50m
Note: Run 25 meters to one side and 25 meters back to the starting position.

3 - Plank: 20 sec.

1 - Abs: 20 times

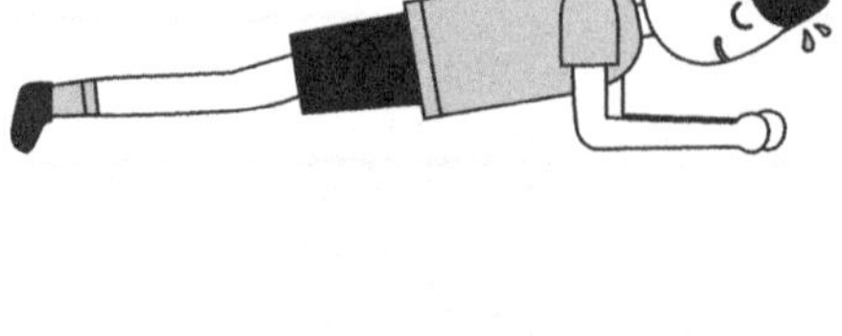

READING PASSAGE

From "Narrative of the Life of Frederick Douglass, An American Slave"

By Frederick Douglass

*(**NOTE:** Continued from Day 1's Passage)*
I had the strongest desire to see Baltimore. Cousin Tom, though not fluent in speech, had inspired me with that desire by his eloquent description of the place. I could never point out any thing at the Great House, no matter how beautiful or powerful, but that he had seen something at Baltimore far exceeding, both in beauty and strength, the object which I pointed out to him. Even the Great House itself, with all its pictures, was far inferior to many buildings in Baltimore. So strong was my desire, that I thought a gratification of it would fully compensate for whatever loss of comforts I should sustain by the exchange. I left without a regret, and with the highest hopes of future happiness.

We sailed out of Miles River for Baltimore on a Saturday morning. I remember only the day of the week, for at that time I had no knowledge of the days of the month, nor the months of the year. On setting sail, I walked aft, and gave to Colonel Lloyd's plantation what I hoped would be the last look. I then placed myself in the bows of the sloop, and there spent the remainder of the day in looking ahead, interesting myself in what was in the distance rather than in things near by or behind.

In the afternoon of that day, we reached Annapolis, the capital of the State. We stopped but a few moments, so that I had no time to go on shore. It was the first large town that I had ever seen, and though it would look small compared with some of our New England factory villages, I thought it a wonderful place for its size — more imposing even than the Great House Farm!

We arrived at Baltimore early on Sunday morning, landing at Smith's Wharf, not far from Bowley's Wharf. We had on board the sloop a large flock of sheep; and after aiding in driving them to the slaughterhouse of Mr. Curtis on Louden Slater's Hill, I was conducted by Rich, one of the hands belonging on board of the sloop, to my new home in Alliciana Street, near Mr. Gardner's ship-yard, on Fells Point.

Mr. and Mrs. Auld were both at home, and met me at the door with their little son Thomas, to take care of whom I had been given. And here I saw what I had never seen before; it was a white face beaming with the most kindly emotions; it was the face of my new mistress, Sophia Auld. I wish I could describe the rapture that flashed through my soul as I beheld it. It was a new and strange sight to me, brightening up my pathway with the light of happiness. Little Thomas was told, there was his Freddy, — and I was told to take care of little Thomas; and thus I entered upon the duties of my new home with the most cheering prospect ahead.

I look upon my departure from Colonel Lloyd's plantation as one of the most interesting events of my life.

WEEK 12 DAY 2

OVERVIEW OF ENGLISH CONCEPTS
READING PASSAGE

1. How was **Cousin Tom** important to Frederick's perceptions and expectations of **Baltimore**?

2. Why does Frederick have some difficulty describing the **precise dates and details** of his journey to Baltimore?

3. What impresses Frederick about **Annapolis**?

   A. The way black and white people interact differently
   B. The wealth of the citizens
   C. The huge and successful shipyards
   D. Its size

4. What does Frederick have to do **first** when he gets to Baltimore?

   A. Meet his new mistress
   B. Meet Little Thomas
   C. Clean himself again
   D. Drive a flock of sheep to the slaughterhouse

5. Based on this passage, how does Frederick's new situation seem **much different** from what he left behind on Col. Lloyd's plantation?

Evaluating a Weak Text

Directions: Think about a text you've read that you thought was **really bad or boring**. It could be a book, a short story, a movie, an episode of a TV show, etc. You could use an **informational text** or a **literary text** for this activity! Write the title of that text on the line below, then answer the following questions to help you think and analyze what makes that text so **bad or boring**!

Text Title: ____________________

- Just generally, what didn't you like about this text?

____

____

- Were there any places in which the text was **unclear or confusing**? Did you ever find yourself frustrated over **feeling lost** in the text?

____

____

____

____

- Why did the text **fail to gain or hold your attention**? Why couldn't you invest in the text? What caused you to tune out or lose interest?

____

____

____

____

OVERVIEW OF ENGLISH CONCEPTS
ACTIVITIES

- Did you feel like there were **applicable takeaways** in this text that connected to your life? If so, why weren't they enough to make you enjoy the text? If not, what could the text have taught or shown you to make things more interesting?

- How could the author have engaged your **intellectual curiosity** better? What might've pushed you to think harder and deeper about the ideas the text was exploring?

WEEK 12
DAY 2

OVERVIEW OF ENGLISH CONCEPTS
ACTIVITIES

- **Why** is this text a **bad or weak example of its genre**? What **stale** or **boring** techniques does it use?

____

____

____

____

____

____

FITNESS

Please be aware of your environment and be safe at all times. If you cannot do an exercise, just try your best.

Repeat these exercises **4 ROUNDS**

1 - Squats: 20 times.
Note: imagine you are trying to sit on a chair.

2 - Side Bending: 15 times to each side. Note: try to touch your feet.

3 - Tree Pose:
Stay as long as possible.
Note: do the same with the other leg.

WEEK 12 DAY 3 MATH

Review

Directions: Use 3.14 as the value for π.

1. What is the area of a rectangular football field 250 ft long by 112 feet wide?

   A. 28,000
   B. 14,000
   C. 11862
   D. 23724

2. What is the area of a rectangular pool with a base of 18 feet and a length of 30 feet?

   A. 263
   B. 116
   C. 270
   D. 540

3. What is the area of the square infield of a baseball diamond that is 85 feet on each side?

   A. 7225
   B. 340
   C. 680
   D. 2720

4. What is the area of a triangular garden with a base of 12 m and a height of 8 m?

   A. 48
   B. 44
   C. 40
   D. 36

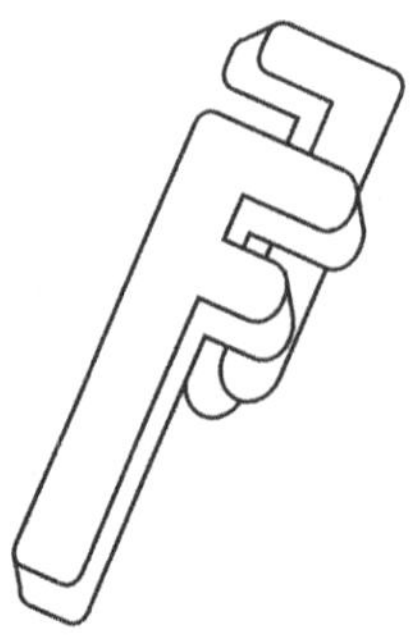

5. What is the approximate area of a circular pool that is 36 yards in diameter?

   A. 1081
   B. 1017
   C. 1170
   D. 1710

6. What is the approximate area of a fountain with a radius of 8 feet?

   A. 194
   B. 197
   C. 201
   D. 205

7. What is the surface area of a can of soup with a diameter of 4 inches and a height of 6 inches?

   A. 73.5
   B. 84.5
   C. 98.5
   D. 100.5

8. What is the surface area of a sugar cube with a side length of 0.75 inches?

   A. 3.38
   B. 13.38
   C. 23.38
   D. 33.38

9. What is the surface area of a ball with a diameter of 7 inches?

   A. 126.9
   B. 153.9
   C. 241.9
   D. 316.9

WEEK 12 DAY 3

MATH

10. What is the surface area of a rectangular prism with a height of 17 meters, a width of 12 meters, and a depth of 6 meters?

A. 756 **C.** 657
B. 765 **D.** 675

11. What is the volume of a court shaped like a rectangular prism with a width of 25 feet, a length of 45 feet, and a height of 12 feet?

A. 10500 **C.** 13500
B. 11500 **D.** 12500

12. What is the volume of a hockey puck with a diameter of $2\frac{1}{2}$ inches and a height of 0.75 inches?

A. 1.32 **C.** 3.68
B. 2.74 **D.** 4.21

13. What is the approximate volume of a globe with a diameter of 18 cm?

A. 1054 **C.** 4054
B. 3052 **D.** 2054

14. What is the approximate volume of a cylinder with a height of 10m and a radius of 8 meters?

A. 2010 **C.** 2013
B. 2012 **D.** 2014

15. What is the volume of a cube with a length of 4 cm?

A. 14 **C.** 16
B. 12 **D.** 64

FITNESS

Please be aware of your environment and be safe at all times. If you cannot do an exercise, just try your best.

Repeat these exercises **4 ROUNDS**

1 - Bend forward: 15 times.
Note: try to touch your feet. Make sure to keep your back straight and if needed you can bend your knees.

2 - Lunges: 12 times to each leg.
Note: Use your body weight or books as weight to do leg lunges.

3 - Plank: 20 sec.

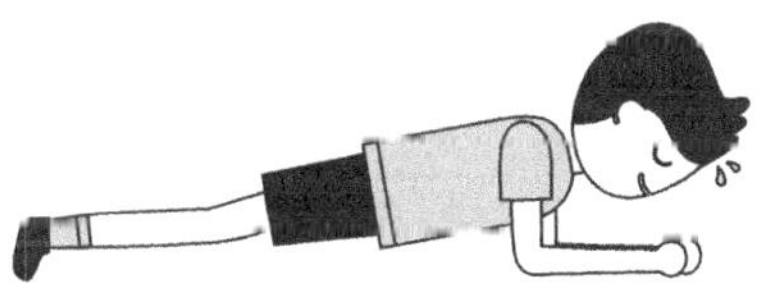

4 - Abs: 20 times

WEEK 12 DAY 4 MATH

Review

1. You have letters that spell M-I-D-D-L-E-S-C-H-O-O-L, what is the probability that you draw an O?

   A. $\frac{1}{3}$ C. $\frac{1}{5}$
   B. $\frac{1}{4}$ D. $\frac{1}{6}$

2. You have letters that spell M-I-D-D-L-E-S-C-H-O-O-L, what is the probability that you draw an S then D?

   A. $\frac{1}{10}$ C. $\frac{1}{9}$
   B. $\frac{1}{66}$ D. $\frac{1}{100}$

3. Simplify ${}_{10}C_4$.

   A. 230 C. 210
   B. 220 D. 200

4. Simplify 6!

   A. 740 C. 700
   B. 720 D. 680

5. Simplify ${}_6P_2$.

   A. 24 C. 28
   B. 26 D. 30

6. What is the probability that you roll an even number on a 6 side die?

   A. $\frac{1}{6}$ C. $\frac{1}{4}$
   B. $\frac{1}{2}$ D. $\frac{1}{3}$

7. What is the probability that you roll a 3 and then a number less than 3 if you roll a 6 sided die twice?

   A. $\frac{1}{18}$ C. $\frac{1}{5}$
   B. $\frac{1}{6}$ D. $\frac{1}{4}$

8. If you take 4 out of 7 shirts on vacation, how many possibilities of shirts do you have?

   A. 760 C. 840
   B. 800 D. 880

9. In a class of 40 students, 13 missed at least 2 days of school last year. Estimate how many students missed that amount in a larger group of 600 students.

   A. 185 C. 205
   B. 195 D. 215

10. There is a rack of 10 bikes on the sidewalk. There are 3 red bikes, 4 blue bikes and 3 white bikes. What is the probability that the next owner that comes will pick a blue bike?

   A. $\frac{5}{10}$ C. $\frac{3}{10}$
   B. $\frac{2}{10}$ D. $\frac{2}{5}$

WEEK 12 DAY 4 MATH

11. If there were 60 bikes on the rack, how many would you expect to be white?

A. 16
B. 18
C. 20
D. 22

12. What is the probability that the next rider will own a green bike?

A. $\frac{0}{10}$
B. $\frac{1}{10}$
C. $\frac{2}{10}$
D. $\frac{3}{10}$

13. A student has 4 white shirts and 4 black shirts. What is probability he will pick a red shirt?

A. 1
B. 0
C. 8
D. 4

14. A student has 4 white shirts and 4 black shirts. They all have buttons. What is the probability he will pick a buttoned shirt?

A. 0%
B. 25%
C. 50%
D. 100%

15. A manufacturer checks 12 cell phones out of a batch of 400 and finds 2 to be defective. How many in the batch can we estimate are defective?

A. 61
B. 64
C. 67
D. 71

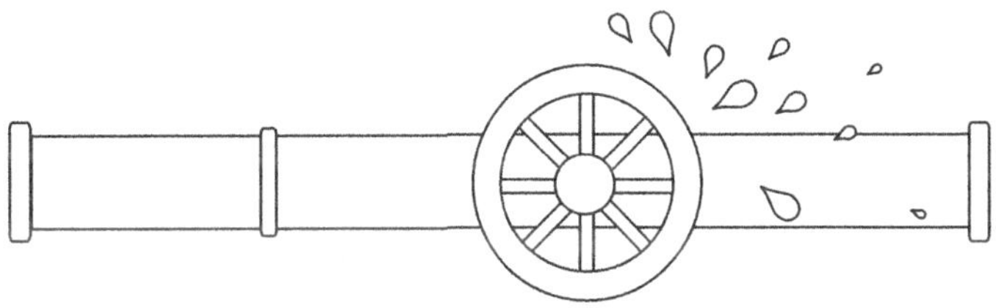

FITNESS

Please be aware of your environment and be safe at all times. If you cannot do an exercise, just try your best.

Repeat these exercises **4 ROUNDS**

1 - High Plank: 25 sec.

2 - Chair: 20 sec.
Note: sit on an imaginary chair, keep your back straight.

3 - Waist Hooping: 20 times. Note: if you do not have a hoop, pretend you have an imaginary hoop and rotate your hips 10 times.

4 - Abs: 25 times

MATH

Review

1. $3x + 4 = 16$

   A. $x = 2$
   B. $x = 3$
   C. $x = 4$
   D. $x = 5$

2. $2y - 12 = 10$

   A. $y = 8$
   B. $y = 9$
   C. $y = 10$
   D. $y = 11$

3. $\frac{4}{7}b + 11 = 23$

   A. $b = 4$
   B. $b = 7$
   C. $b = 14$
   D. $b = 21$

4. $9z - 6 = 102$

   A. $z = 16$
   B. $z = 14$
   C. $z = 12$
   D. $z = 10$

5. $14n + 2 = 142$

   A. $n = 8$
   B. $n = 10$
   C. $n = 12$
   D. $n = 14$

6. $\frac{w}{7} + 11 = 22$

   A. $w = 77$
   B. $w = 30$
   C. $w = 27$
   D. $w = 24$

7. $18 - \frac{e}{7} = 9$

   A. $e = 27$
   B. $e = 36$
   C. $e = 45$
   D. $e = 63$

8. $10 + 4n = 38$

   A. $n = 4$
   B. $n = 5$
   C. $n = 6$
   D. $n = 7$

9. $22 + 6x = 34$

   A. $x = 4$
   B. $x = 3$
   C. $x = 2$
   D. $x = 1$

10. $-4y + 2 = 10$

    A. $y = -1$
    B. $y = -2$
    C. $y = -3$
    D. $y = -4$

WEEK 12 DAY 5 MATH

11. $7u - 9 > 40$

A. $u > 7$
B. $u > 6$
C. $u > 5$
D. $u > 4$

12. $15t + 2 < 62$

A. $t < 3$
B. $t < 4$
C. $t < 5$
D. $t < 6$

13. $11k - 3 > 217$

A. $k > 14$
B. $k > 16$
C. $k > 18$
D. $k > 20$

14. $13n + 3 < 120$

A. $n < 7$
B. $n < 8$
C. $n < 9$
D. $n < 10$

15. $-4n + 2 > 38$

A. $n > 9$
B. $n < 9$
C. $n < -9$
D. $n > -9$

YOGA

Please be aware of your environment and be safe at all times. If you cannot do an exercise, just try your best.

1 - Down Dog: 30 sec.

2 - Bend Down: 30 sec.

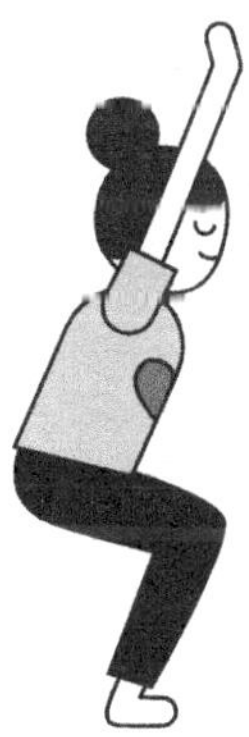

3 - Chair: 30 sec.

4 - Child Pose: 30 sec.

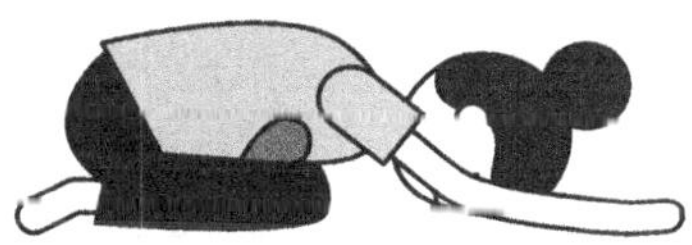

5 - Shavasana: as long as you can. Note: think of happy moments and relax your mind.

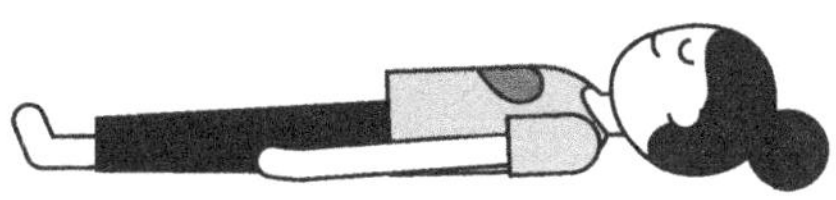

WEEK 12 DAY 6

EXPERIMENT

Traits from Two Parents

From Weeks 8-11, you learned how different **microorganisms** reproduce through various processes. Most cells simply copy their genetic information, creating exact clones, but large, **multicellular** organisms like plants and animals don't just reproduce by cloning! While a cell only has one "parent," plants and animals typically have **two** genetic parents. That means that those organisms are made up from two combined DNA sequences.

When genetic sequences are combined, a completely new roadmap is created from two different sets of instructions. That's why people frequently display strong **traits** from both of their genetic parents, while also, in some ways, expressing a **combination** of their ancestors' qualities. The 19th century monk Gregor Mendel was one of the first scientists to figure out how this process worked by studying pea plants.

Today, you'll do an activity that mirrors Mendel's work and primes you to think about the inheritance of traits.

Materials:

- Graph paper
- A ruler
- Art supplies (colored pencils, markers, etc.)
- Internet access (for research)
- Notepaper

EXPERIMENT

Procedure:

1. Using internet research, define the terms **Dominant Trait** and **Recessive Trait** in your notes. If you want, you can also research **Gregor Mendel** to learn more about his work with pea plants.

2. Using internet research, look up the term **Punnett Square** and take notes on how Punnett squares are used to study and examine gene inheritance and expression.

3. Using your ruler and graph paper, draw 4 different Punnett Squares. We will be imagining we are studying **pea plants**, just like Gregor Mendel.

4. Label your first Punnett square "**POD COLOR.**" Label the top (y-axis): "**YELLOW - YY.**" Label the side (x-axis): "**GREEN - yy.**" Once your labeling is done, complete the Punnett square to determine what colors the pods of four offspring of those two pea plants would be. Next to each of your results, use your art supplies to draw a small picture of what the results might look like.

5. Label your second Punnett square "**FLOWER COLOR.**" Label the top (y-axis): "**WHITE - ww.**" Label the side (x-axis): "**PINK - Ww.**" Once your labeling is done, complete the Punnett square to determine what colors the flowers of four offspring of those two pea plants would be. Next to each of your results, use your art supplies to draw a small picture of what the results might look like.

6. Label your third Punnett square "**SEED SHAPE.**" Label the top (y-axis): "**ROUND" - Rr.**" Label the side (x-axis): "**OBLONG - rr.**" Once your labeling is done, complete the Punnett square to determine what shape the seeds of four offspring of those two pea plants would be. Next to each of your results, use your art supplies to draw a small picture of what the results might look like.

7. Label your fourth Punnett square "**POD TEXTURE.**" Label the top (y-axis): "**SMOOTH" - SS.**" Label the side (x-axis): "**BUMPY - Ss.**" Once your labeling is done, complete the Punnett square to determine what texture the pods of four offspring of those two pea plants would be. Next to each of your results, use your art supplies to draw a small picture of what the results might look like.

8. Clean up your materials and answer the questions below.

WEEK 12 DAY 6 EXPERIMENT

Follow-Up Questions:

1. Based on what you saw with your Punnett squares, why are certain traits represented over and over again in certain families?

____

____

____

2. Based on what you saw with your Punnett squares, how can people be carriers for traits they don't even necessarily express?

____

____

____

YOGA

Please be aware of your environment and be safe at all times. If you cannot do an exercise, just try your best.

2 - Down Dog: 30 sec.

3 - Stretching: Stay as long as possible.
Note: do on one leg then on another.

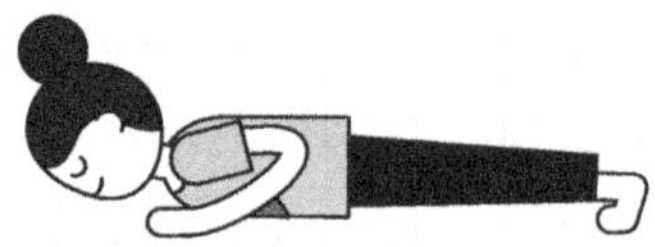

4 - Lower Plank: 20 sec.
Note: Keep your back straight and body tight.

1 - Tree Pose: Stay as long as possible.
Note: do on one leg then on another.

6 - Shavasana: 5 min.
Note: this pose is very important and provides you with long term benefits. Try not to skip this. Close your eyes and imagine who you want to be and what your goals are! Always think happy thoughts.

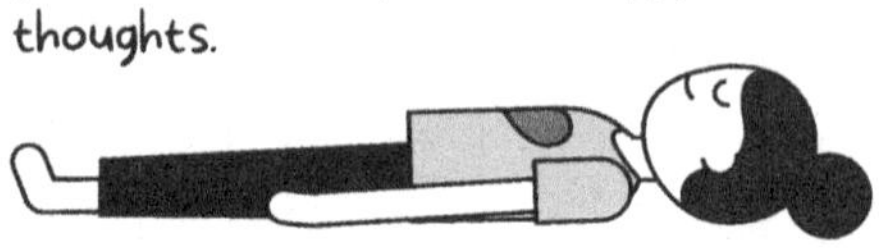

5 -Book Pose: 20 sec.
Note: Keep your core tight. Legs should be across from your eyes.

WEEK 12
DAY 7

GOALS

Task: Write down 10 goals you would like to accomplish this year.

GOAL #1

GOAL #2

GOAL #3

GOAL #4

GOAL #5

GOAL #6

GOAL #7

GOAL #8

GOAL #9

GOAL #10

Answer Sheets

To see the answer key to the entire workbook, you can easily download the answer key from our website!

*Due to the high request from parents and teachers, we have removed the answer key from the workbook so you do not need to rip out the answer key while students work on the workbook.

Go to **argoprep.com/summer8**
OR scan the QR Code:

Download Answers
Kids Summer Academy by ArgoPrep: Grade 8-9
Kids Summer Academy by ArgoPrep: Grade 5-6
Kids Summer Academy by ArgoPrep: Grade 4-5
Kids Summer Academy by ArgoPrep: Grade 6-7

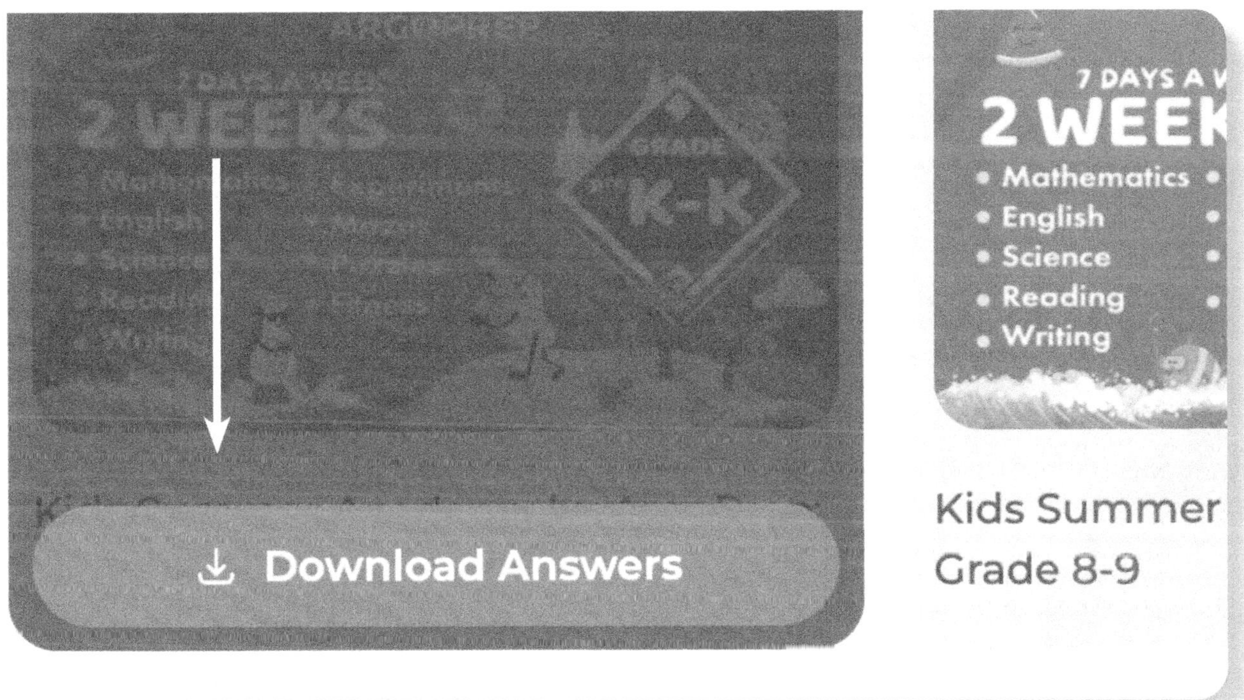
Download Answers
7 DAYS A W
2 WEEK
Mathematics
English
Science
Reading
Writing
Kids Summer
Grade 8-9

Made in the USA
Las Vegas, NV
11 April 2024

88495453R00149